AF239541

Eda Özden

Discrete Time Analysis of Consolidated Transport Processes

Wissenschaftliche Berichte des
Institutes für Fördertechnik und Logistiksysteme
des Karlsruher Instituts für Technologie
Band 77

Discrete Time Analysis of Consolidated Transport Processes

by
Eda Özden

Dissertation, Karlsruher Institut für Technologie
Fakultät für Maschinenbau, 2011
Referent: Prof. Dr.-Ing. K. Furmans
Korreferentin: Prof. Dr. K.-H. Waldmann

Impressum

Karlsruher Institut für Technologie (KIT)
KIT Scientific Publishing
Straße am Forum 2
D-76131 Karlsruhe
www.ksp.kit.edu

KIT – Universität des Landes Baden-Württemberg und nationales
Forschungszentrum in der Helmholtz-Gemeinschaft

KIT Scientific Publishing 2012
Print on Demand

ISSN 0171-2772
ISBN 978-3-86644-801-8

Discrete Time Analysis of Consolidated Transport Processes

Zur Erlangung des akademischen Grades eines

Doktors der Ingenieurwissenschaften

von der Fakultät für Maschinenbau
des Karlsruher Instituts für Technologie (KIT)
genehmigte

Dissertation

von

MSc Eda Özden

aus Istanbul

Tag der mündlichen Prüfung:	13. Juli 2011
Hauptreferent:	Prof. Dr.-Ing. K. Furmans
Korreferent:	Prof. Dr. K.-H. Waldmann

Vorwort

Die vorliegende Arbeit entstand während meiner Tätigkeit als wissenschaftliche Mitarbeiterin am Institut für Fördertechnik und Logistiksysteme (IFL) des Karlsruher Instituts für Technologie (KIT). Dieses Vorwort möchte ich nutzen, um denjenigen zu danken, die zum Gelingen der vorliegenden Dissertationsschrift beigetragen haben.

Mein besonderer Dank gilt Herrn Prof. Dr. -Ing. Kai Furmans, dem Leiter des Instituts für Fördertechnik und Logistiksysteme, für die Übernahme des Hauptreferats, die vielen fachlichen Anregungen sowie die Unterstützung während meiner Zeit am Lehrstuhl. Seine Forschungsarbeiten im Bereich der bedientheoretischen Methoden haben mich zu dieser Arbeit motiviert.

Herrn Prof. Dr. Karl-Heinz Waldmann, dem Leiter des Instituts für Wirtschaftstheorie und Operations Research des KIT, danke ich für die Übernahme des Korreferats und besonders für die anregenden Diskussionen. Die daraus hervorgehenden Impulse haben die Arbeit stets bereichert. Weiterhin danke ich Herrn Prof. Dr.-Ing. Carsten Proppe für die Übernahme des Prüfungsvorsitzes.

Bedanken möchte ich mich auch beim Karlsruhe House of Young Scientists für die finanzielle Unterstützung durch ein Promotionsstipendium.

Meinen Kollegen danke ich für die sehr freundliche Zusammenarbeit. Besonderer Dank gebührt Judith Stoll für die konstruktiven Ratschläge.

Mein tiefster persönlicher Dank gilt meinen Eltern und Schwestern sowie Robert. Das Gelingen dieser Arbeit habe ich im allerhöchsten Maße ihrer Zuneigung zu verdanken.

Karlsruhe, Juli 2011   Eda Özden

Kurzfassung

Eda Özden
Zeitdiskrete Analyse von Transportbündelungen

Ziel der vorliegenden Arbeit ist die Entwicklung zeitdiskreter Modelle zur Analyse von Transportbündelungen. Die zeitdiskrete Modellierung ermöglicht eine Beschreibung von Leistungskenngrößen mit Hilfe von generellen stochastischen Verteilungen, im Gegensatz zur zeitkontinuierlichen Modellierung, die eine Beschreibung durch Mittelwerte und Varianzen zugrunde legt. Dadurch wird ermöglicht, die in der Logistik häufig benötigten Quantile von Kennwerten zu bestimmen. Basierend auf Quantilen können Puffer in einem Transportsystem so ausgelegt werden, dass sie z.b. in 95% der Fälle für ankommende Transporteinheiten ausreichend dimensioniert sind.

Zunächst werden die Bestandsbündelungen untersucht, bei welchen die Lieferungen hinausgezögert werden. Dies kann andauern bis ein bestimmtes Transportlos erreicht, oder eine vordefinierte Zeitspanne vergangen ist. Zu diesem Zweck werden zwei verschiedene Batchbildungsmodelle analysiert. Im ersten Modell erfolgt der Transport der gesammelten Transporteinheiten entweder, wenn eine maximale Sammelzeit vergangen ist, oder, wenn die Kapazität des Transportfahrzeugs erschöpft ist. Im Unterschied dazu wird im zweiten Modell ein Transportprozess ausgelöst, sobald eine vorgegebene Mindestauslastung für das eingesetzte Fahrzeug erreicht wird. Für beide Modelle werden Wartezeit- und Zwischenabgangszeitverteilungen exakt bestimmt. Zusätzlich wird das $G^X/G^{[L,K]}/1$-Bediensystem analysiert. Auch für dieses Modell werden Wartezeit- und Zwischenabgangszeitverteilungen bestimmt. Die Ergebnisse sind exakt in einer ϵ-Umgebung.

Die Fahrzeugbündelung beschreibt den Fall, in dem Fahrzeuge auf einer Tour an mehreren Stationen Ladung aufnehmen oder abliefern. Für die Analyse von Fahrzeugbündelungen werden Modelle für getaktete und shuttle Milkrun-Systeme entwickelt. Mit den vorgestellten Modellen ist es möglich, einen Materialfluss in beide Richtungen abzubilden, indem nicht nur der Fluss der Materialien selbst gezeigt wird, sondern es kann zum Beispiel ebenfalls die Bewegung von leeren Behältern in das Modell integriert werden. Für beide Systeme werden iterative Algorithmen entwickelt, die die Verteilungen der Systemzustände, Tour-Dauer, Zykluszeit und der Wartezeit approximieren.

Abschließend wird die Umschlagslagerbündelung analysiert, in der Ware indirekt über Umschlagspunkte zum Ziel transportiert wird. Für die Umschlagslagerbündelung wird eine Netzwerkanalyse durchgeführt.

Abstract

Eda Özden
Discrete Time Analysis Of Consolidated Transport Processes

The objective of this work is to develop discrete time models for the analysis of consolidated transport processes. In contrary to continuous time models, discrete time models enable an analysis of key performance measures on the basis of general distributions instead of mean values and variances. As a result, it is possible to compute quantiles of performance measures, which are often needed in the field of logistics. On the basis of computed quantiles, buffer sizes in a transport system can be so determined that incoming transport units can be accommodated with a given probability (e.g. 95%).

Firstly, inventory consolidation is studied, under which deliveries are delayed until either a predefined transport quantity is reached or a given time interval is elapsed. For this purpose, two batch building models are studied. In the first model, collected batches are transported when either vehicle capacity or a given maximum collecting time is reached. On the other hand, collecting process finishes under the second model, as soon as a predefined minimum utilization of the vehicle is achieved. For both models, exact distributions of inter-departure time, departing batch size, and waiting time are derived. Moreover, the $G^X/G^{[L,K]}/1$-queue is investigated. A detailed analysis of the waiting time and inter-departure time is provided. Results are exact within an ϵ-neighborhood.

The next consolidation strategy studied is vehicle consolidation, which is characterized by employing the same vehicle to serve a couple of receiving and/or shipping points successively. For the analysis of vehicle consolidation, stochastic models for taktet and shuttle milkrun systems are developed. In both models, a bidirectional material is allowed by considering not only the flow of goods, but also the flow of e.g. empty containers. Iterative algorithms are developed for both models, which compute cycle time, queue states, tour time, and waiting time distributions approximately.

Finally, for terminal consolidation, under which material streams are transported to their destinations over transit terminals, a network analysis is conducted.

Contents

Kurzfassung   iii

Abstract   v

1. Introduction   1
   1.1. Problem Description and Scope of the Book   3
   1.2. Organization of the Book   4

2. Discrete Time Queuing Analysis   7
   2.1. Basic Definitions in Discrete Time Domain   9
   2.2. Basic Operations in Discrete Time Domain   11
   2.3. Notation Principles   12

3. Queuing Analysis of Consolidated Transport Processes   13
   3.1. Literature Review on Batch Processes   13
      3.1.1. Batch Arrival Queue   14
      3.1.2. Batch Server Queue   16
      3.1.3. Batch Queue   17
      3.1.4. Batch Building Processes   19
   3.2. Literature Review on Polling Systems   20
      3.2.1. Discrete Time Polling Systems   28

4. Inventory Consolidation   31
   4.1. Introduction   31
   4.2. Batch Building: Capacitated Timeout Rule   32
      4.2.1. Queuing System   32
      4.2.2. Inter-departure Time Distribution   39
      4.2.3. Departing Batch Size Distribution   40
      4.2.4. Waiting Time Distribution   41
      4.2.5. Numerical Results   46

4.3. Batch Building: Capacity Interval Rule 48
    4.3.1. Queuing System 49
    4.3.2. Departing Batch Size Distribution 52
    4.3.3. Inter-departure Time Distribution 53
    4.3.4. Waiting Time Distribution 53
4.4. Batch Queue: $G^X/G^{[L,K]}/1$-Queue 55
    4.4.1. Queuing System 56
    4.4.2. Departing Batch Size Distribution 62
    4.4.3. Inter-departure Time Distribution 64
    4.4.4. Waiting Time Distribution 65
    4.4.5. Numerical Results 70

5. Vehicle Consolidation   **79**
5.1. Takted Milkrun Systems 81
    5.1.1. Queuing System 83
    5.1.2. Iterative Algorithm 85
    5.1.3. Cycle Segment Distributions 86
    5.1.4. Queue State Distributions 89
    5.1.5. Tour Time Distribution 93
    5.1.6. Cycle Time Distributions 93
    5.1.7. Waiting Time Distributions 93
    5.1.8. Improvement Algorithm 95
    5.1.9. Numerical Results 108
5.2. Shuttle Milkrun Systems 116
    5.2.1. Queuing System 116
    5.2.2. Iterative Algorithm 118
    5.2.3. Queue State Distributions 121
    5.2.4. Cycle and Tour Time Distributions 122
    5.2.5. Waiting Time Distributions 125
    5.2.6. Numerical Results 125

6. Terminal Consolidation   **129**
6.1. Introduction 129
6.2. Numerical Case: Analysis of a Hub-and-Spoke Network   131

7. Conclusion   **139**

8. Glossary   **145**

Bibliography   **149**

A. Appendix   **159**
    A.1. Analysis of the $G^X/G^{[L,K]}/1$-queue   159
    A.2. Analysis of the takted milkrun systems   162
    A.3. Analysis of the shuttle milkrun systems   165
    A.4. Numerical case study   170

1. Introduction

Ours is a planet of continuous improvement and innovation; humans are creative thinkers and restless explorers limited only by imagination. One of the sectors that benefited most from mankind's drive to innovate is the transport sector, of which history is marked with several evolutionary steps. All these evolutionary steps, from the start of man-made transport with the invention of wheel till man's travel to the moon, have opened new opportunities. With the invention of wheel, mankind developed the idea of mass transport and the idea of trade and exchange of goods. And after Wright brothers invented the first plane, long distances became less important in our lives. Although we are still not quite sure about the level of innovation, if there is any, in other planets, we leave this question to the next generations and move onto the current issues in transport systems.

Contemporary transport systems supported the expansion of the global economy by enabling movement of passengers and freight. However, globalism generates major challenges for the transport systems. Firstly, existing transport systems experience problems to keep pace with increasing customer wishes for higher reliability, more product diversification, lower unit prices, higher flexibility, faster delivery etc. At the same time, transport volumes grow rapidly. As the result, traffic congestion causes delay and unreliable deliveries. An immediate solution is to utilize the capacities more efficiently by tying up different transportation modes. Last, but not least, the long trend of growing energy costs and environmental concerns becomes more serious and imposes significant adjustments to the transport systems. In particular, environmentally friendly and energy efficient transportation modes will be employed more intensively. In the light of these challenges, existing transport systems must be restructured with innovative technology and management expertise.

In order to overcome these challenges, many logistics concepts have been developed. For instance, the concept of multi-modal freight centers is conceived to bundle a number of transportation modes in a facility near to an urban area. In this way, it is possible to facilitate consolidation of deliveries and promote deployment of more energy efficient transportation modes, e.g. sea transport, rail transport, especially for long distance travels. However, rail and waterways depend still on road transport for final deliveries in this concept. A new concept to decrease the dependency on the road transport is the concept of underground freight transportation. Sending goods in underground pipelines is indeed not new. We already know underground gas pipelines or pneumatic post networks in some large buildings. But the idea of implementing the concept for solid freight is new. Countries like Germany and Holland with heavy traffic congestion focus on developing such a concept, which involve unmanned electric vehicles on rails that travel through pipelines. Due to the efficient electric drive, the concept is more energy efficient and requires less space. Thus, such a concept may evolve in future to the fifth transportation mode next to road, rail, sea, and air transports. Not only external transport systems, but also internal transport systems are subject to change. An example is the replacement of bulky conveying systems by flexible automated guided vehicle (AGV) systems in intralogistics systems, which fits to today's lean world. In AGV systems, a number of vehicles traverse predefined routes to visit a number of workstations. In contrary to conveyor systems, routes can be redefined easily without touching the infrastructure.

All these concepts reported above have something in common: they all rely on one or more consolidation strategies. We explain consolidation strategies on the example of subways, in which one can see all three strategies. Passengers arrive at stations and wait until the next tram arrives. Thus, the tram picks up a group of people that accumulated at each station. This kind of consolidation is called inventory consolidation. After that the tram drives to the next station, where it drops off a number of passengers and picks new ones. In subways, trams visit a number of stations on their routes. The idea of employing the same vehicle to serve multiple stations constitutes vehicle consolidation. In such a network of stations, some stations are declared to be main stations, through which main transport streams flow. Passengers are routed to the nearest of the main stations, where they change to another tram to move to their final destinations. This is called terminal consolidation. The opposite of the consolidated transport processes is direct transport. This is the case, when the passenger decides to travel to work by his private car. So the person accepts to incur more costs in exchange for more flexibility, shorter travel times, more comfort, better access etc.

The idea behind consolidation strategies is to utilize existing transport capacities more efficiently. In this way, it is possible to lower unit operational costs owing to economies of scale. However, they all increase sojourn time in some ways, which may yield a deterioration in flexibility and service level. Inventory consolidation increases delay time and vehicle consolidation increases travel time. Terminal consolidation increases both travel and handling time. Furthermore, it requires considerable investment in infrastructure. Hence, effects of consolidation must be quantified. The objective of the current work is to develop analytical approaches for the analysis of consolidated transport processes.

1.1. Problem Description and Scope of the Book

For an efficient design of transport systems, effects of consolidation, degree of consolidation, and system parameters should be analyzed. Consider the design process of an underground freight system. The planner has to answer numerous questions before a prototype is generated. What should be the capacity of the electric vehicles? How fast should the vehicles move? Or are cost benefits sufficient to cover substantial investment needed? etc. Or suppose that an important change in the production technology happened, leading to a change in the route of an AGV-system. Then, the question to answer is how the throughput of the system changes with the new route design. Or alternatively, the planner may need to resize the puffers at visited stations. Regarding multi-modular freight centers, one can question, which improvements in utilization can be achieved, or how the sojourn time changes. Obviously, many of these design questions cannot be answered simply based on mean values. For instance, a puffer or a terminal is dimensioned based a required safety level. Similarly, determination of the mean sojourn time would not help to determine the probability of on-time order fulfillment. Thus, planners need efficient methods, that deliver adequate level of detail.

In this work, we are motivated to develop appropriate models for the analysis of different consolidation strategies. We develop analytical models for different inventory and vehicle consolidation strategies. Besides, we show how these models can be used together to analyze terminal consolidation strategies. Eventually, we derive insights into application related aspects of the presented analytical approaches.

We assume in our models, that time is discrete, as this offers many advantages. Discrete time approaches are most suited to be employed in an early planning phase. The reasons for this are computational efficiency and adequate level of detail offered by discrete time analysis. In an early planning phase, mostly rough data or approximations exist. So a great deal of sensitivity analysis is needed. Moreover, different configurations for system parameters have to be studied. With discrete time analysis, it is possible to study a number of scenarios within a short time. The level of detail is sufficient to support strategic planning, as it allows an analysis based on distributions, not only on mean values. Concluding, the objectives of this work include:

- Development of analytical approaches for the evaluation of consolidated transport systems,
- Analysis of transport processes in a stochastic setting,
- Discussion of application related aspects,
- Decision-making support for the (re-)design of transport systems in an early planning phase.

1.2. Organization of the Book

This work is organized as follows: in chapter 2, discrete time queuing analysis is introduced. At first, discrete time theory, its assumptions as well as its advantages and disadvantages are explained. Afterwards, basic definitions of discrete time probability theory are provided. Moreover, a number of standard operators used in our analysis is presented.

Finally, the notation system used for the mathematical description of the models is explained. In Chapter 3, fields of queuing theory, which are related to this work, are introduced. The first field is queuing analysis of batch processes, where we further distinguish between batch arrival queues, batch server queues, batch queues, and batch building processes. Subsequently, the literature on batch processes is reviewed following this classification. The second field investigated is polling systems (also known as cyclic server systems). For this queuing discipline, a structured summary of general polling models is given. Subsequently, the literature on discrete time polling systems is discussed. In chapter 4, inventory consolidation is studied. At first batch building processes under capacitated timeout and capacity interval rules are analyzed. For both rules, exact distributions of inter-departure time, departing batch size and waiting time are derived. Thereafter, we study the $G^X/G^{[L,K]}/1$-queue. In this analysis, a two dimensional Markov-Chain is used to derive the performance measures. The method is exact within an ϵ-neighborhood. Finally, it is shown how introduced methods can be used to evaluate different inventory consolidation strategies. Chapter 5 is devoted to vehicle consolidation. Specifically, takted and shuttle milkrun systems are investigated. For these systems, cycle time, queue states, tour time, and waiting time distributions are computed approximately. For takted milkrun systems, we develop an improvement algorithm that can be used to correct the results of the basic algorithm, when the material flow in the given system is deterministic. Otherwise, the basic algorithm delivers sufficiently accurate results. Chapter 6 discusses terminal consolidation by analyzing a hub-and-spoke transport network. For this purpose, we build up the network with different model elements. Eventually, we compute the sojourn time for two streams and compare them with simulation results.

2. Discrete Time Queuing Analysis

In discrete time analysis, time is partitioned into intervals of unit length (Δt) and discrete probability environments are used to define input variables. Therefore, e.g. inter-arrival time between incoming customers at a discrete time queue must be a multiple of Δt. Consequently, events occur in discrete time models only at boundary epochs of unit intervals, simplifying modeling efforts.

Over the last two decades, there has been an increased interest in discrete time analysis, which can be attributed to its various applications in computer and digital telecommunication networks. Specifically, broadband integrated services digital network (B-ISDN), which provides a common interface for the transfer of video, voice, and data, has received considerable attention. The basic transfer mode in B-ISDN is the Asynchronous Transfer Mode (ATM), which transmits the information in small, fixed-length packets. Thus, the system is described in discrete time domain better than in continuous time domain. Besides telecommunication systems, systems in other areas operate increasingly in clocked cycles. A good example for such a system is the assembly line production, in which operations of the assembly stations are synchronized by adhering to a given cycle time. Thus, arrivals and departures of products are allowed only at equally spaced points in time. Discrete time analysis is also suitable for this case.

As explained in the above discussion, some systems are "naturally" discrete and discrete time analysis is the adequate technique for performance analysis of such systems. For some other systems, discrete time analysis can still be the preferred technique as it offers many advantages.

An important advantage of discrete time analysis is related to input data. In discrete time analysis, distributions of input variables are based on measured data. By contrast, input variables are approximated in continuous time domain by a theoretical distribution, most favorably by the exponential distribution, in order to get the benefit of its memoryless property. Provided that the Markovian property can be assumed for some variables, analysis simplifies to a great extend. Since a Markov chain can be embedded. In this case, the quality of the results is affected by the goodness of fit to the theoretical distribution.

However, there are instances, where input variables cannot be fit to a known distribution. When this is the case, the analysis is mostly limited to the information delivered by the first two moments of the input distributions. For the mean waiting time of the $G/G/1$-queue, Schleyer (2007) compares the results of the 2-parameter approximation methods introduced by Whitt (1993) and Bolch et al. (1998) with the discrete time approach by Grassmann and Jain (1989). The results show that the 2-parameter approximation methods may yield remarkably high deviations depending on the variability of the input distributions and the utilization of the system (see Schleyer (2007) for a detailed analysis). In some cases one approximation works better than the other one. In contrary, the discrete method is exact within an ϵ-neighborhood for all cases. In general, discrete time approaches are robust and, hence, applicable to a wide variety of applications.

In many cases, an analysis based on distributions is needed. In order e.g. to size a puffer with a given statistical safety, an analysis based on mean and the variance does not help, as one needs to know percentiles of queue length. Unfortunately, most of the literature in continuous time domain delivers mean and the variance of performance measures and only for simple models percentiles can be computed. In such cases, simulation is used to study the system. However, simulation studies reach their limits quickly regarding necessary computing time. Furthermore, a great deal of time and statistical expertise are needed to validate the model and to evaluate the simulation results.

With discrete time analysis, it is possible to calculate whole distributions of performance variables efficiently. Discrete time models are usually built by using a couple of standardized operations e.g. convolution, pi-operators etc and are easier to understand and to implement (see Tran-Gia and Ahmadi (1988)). These operations can be enumerated efficiently by means of powerful algorithms from the signal processing theory. An example is the Fast Fourier Transform (FFT) used for the convolution operator.

The theory of the discrete time analysis is diverse. Generally, one can differentiate between the analysis techniques that operate in time domain and the techniques that operate in transform domain. In the scope of this thesis, we develop discrete time techniques in time domain, which are easier to understand and to implement compared to the techniques in transform domain. This chapter familiarizes the reader with the basic discrete time queuing theory in time domain. In section 2.1, we introduce the basic definitions of the probability theory in discrete time domain. Subsequently, we present in section 2.2 basic operators, we commonly employ in our methods. Eventually, we make a note on the notation system employed in our analysis. For a more detailed discussion of discrete time theory, interested reader may refer to Schleyer (2007), Furmans (2004a), Furmans (2004b), Tran-Gia (1996), Tran-Gia (1993), and Bruneel and Kim (1992).

2.1. Basic Definitions in Discrete Time Domain

In our analysis, we observe the system at the multiplies of the unit length Δt. Events are described by discrete random variables. Given the discrete random variable (RV) $\mathbf{X}$ and let $0, \Delta t, 2 \cdot \Delta t, \cdots, x_{max} \cdot \Delta t$ be the values, it assumes, we denote its distribution by

$$P(X = i \cdot \Delta t) = P(X = i) = x_i \qquad \forall i : 0, 1, 2, \cdots, x_{max} \tag{2.1}$$

In order to simplify the notation, time measures are normalized on the basis of Δt, e.g. instead of $X = i \cdot \Delta t$, the normalized form $X = i$ is used. The cumulative distribution function of $\mathbf{X}$ is given by

$$P(X \leq i) = \sum_{m=0}^{i} x_m \tag{2.2}$$

9

The mean or the expected value of $\mathbf{X}$ is defined by

$$E(X) = \sum_{i=0}^{x_{max}} i \cdot x_i \tag{2.3}$$

If the expectation of the random variable X^n exists, it is called the nth moment of $\mathbf{X}$. Therefore, we define

$$E(X^n) = \sum_{i=0}^{x_{max}} i^n \cdot x_i \tag{2.4}$$

The variance of the variable is then given by

$$VAR(X) = E((X - E(X))^2) = E(X^2) - E(X)^2 \tag{2.5}$$

Then, the squared coefficient of variation (scv) is denoted by

$$c_X^2 = \frac{VAR(X)}{E(X)^2} \tag{2.6}$$

and the $u\%$ quantile of the variable $\mathbf{X}$ (σ_u) is the value ω, such that the cumulative distribution function is greater than or equal to u.

Consider now two discrete RV $\mathbf{X}$ and $\mathbf{Y}$ defined on the same probability space. The joint probability for the events $(X = i)$ and $(Y = j)$ occurring at the same time is given by $P(X = i \ \wedge \ Y = j)$. Besides, conditional probabilities are used in our analysis. The conditional probability $P(X = i \mid Y = j)$ is the probability that the event $(X = i)$ occurs given the occurrence of $(Y = j)$ and is described by

$$P(X = i \mid Y = j) = (x_i \mid Y = j) = \frac{P(X = i \ \wedge \ Y = j)}{P(Y = i)} \tag{2.7}$$

With the law of total probability, we can obtain the probability $P(X = i)$ based on the conditional probabilities

$$P(X = i) = \sum_{j=0}^{y_{max}} P(X = i \mid Y = j) \cdot P(Y = j) \tag{2.8}$$

Eventually, we explain here briefly the discrete renewal process and give the expressions for the residual life time of a renewal process (see Tran-Gia (1996) and Schleyer (2007) for a detailed discussion). Renewal processes are point processes, in which inter-event times are independent and identically distributed (iid). The name "renewal process" is motivated by the fact, that the process is reset at each occurrence of each event, called the renewal points. If t is an arbitrary time instant, at which the process is observed, the residual life time is defined as the time from t to the occurrence of the next event. As time is discrete, we differentiate between two cases. If the residual lifetime is observed immediately after discrete time instants, its distribution is given by

$$r_i = \frac{1}{E(X)}(1 - \sum_{j=0}^{i-1} x_j) \qquad \forall i : 1, \cdots, x_{max} \tag{2.9}$$

If the observation takes place immediately before discrete time instants, the distribution of the residual life time is defined as

$$r_i = \frac{1}{E(X)}(1 - \sum_{j=0}^{i} x_j) \qquad \forall i : 0, \cdots, x_{max} - 1 \tag{2.10}$$

2.2. Basic Operations in Discrete Time Domain

We explain in this section the basic operations, we use often in our analysis. In order to calculate the distribution of a RV $\mathbf{Z}$, which is sum of independent RV $\mathbf{X}$ and $\mathbf{Y}$, we use the convolution operation, which is defined below

$$z_i = \sum_{-\infty}^{\infty} x_j \cdot y_{i-j} = x \otimes y \tag{2.11}$$

Likewise, we employ the negative convolution, to compute the distribution of a RV Z, which is difference of independent RV X and Y

$$z_i = \sum_{-\infty}^{\infty} x_{i+j} \cdot y_j = x \otimes -y \tag{2.12}$$

Based on the description in Dittmann and Hübner (1993), we use the pi-operators Π_m and Π^M in order to obtain the distribution of a RV $\mathbf{X}$ with a lower bound m or an upper bound M, respectively.

$$\Pi_m[x_i] = \begin{cases} 0 & i < m \\ \sum_{j=-\infty}^{m} x_j & i = m \\ x_i & i > m \end{cases} \qquad \Pi^M[x_i] = \begin{cases} x_i & i < m \\ \sum_{j=M}^{\infty} x_j & i = m \\ 0 & i > m \end{cases}$$

$$(2.13)$$

Finally, we shift the distribution of a RV $\mathbf{X}$ up by m units with the following operator

$$\Delta_m[x_i] = x_{i+m} \qquad (2.14)$$

2.3. Notation Principles

We denote distributions with the associated small letters. Thus, the distribution of a RV $\mathbf{X}$ is given by $x_i = P(X = i)$. We summarize here the notation principles, that we adopted for the mathematical description in this work.

$P(X = i)$ or x_i	probability that RV $\mathbf{X}$ is i units,
$x_j^{l\otimes}$	probability that the l-fold convolution of RV $\mathbf{X}$ with itself is j units,
$\overline{x_i}$	probability that RV $\mathbf{X}$ assumes a value greater than or equal to i units,
x_{min}	minimum value of RV $\mathbf{X}$,
x_{max}	maximum value of RV $\mathbf{X}$,
$E(X), VAR(X)$	expected (mean) value and variance of RV $\mathbf{X}$, respectively,
$P(X = i \wedge Y = j)$	joint probability that events $(X = i)$ and $(Y = j)$ occur simultaneously,
$P(X = i \mid Y = j)$ or $(x_i \mid Y = j)$	conditional probability that RV $\mathbf{X}$ assumes a value of i time units, given $(Y = j)$.

3. Queuing Analysis of Consolidated Transport Processes

Queuing analysis of consolidated transport processes involves, on one hand, queuing analysis of batch processes, on the other hand, polling systems, or also known as cyclic server systems. Section 3.1 gives a brief introduction to the theory of batch processes and then discuss the relevant literature. In section 3.2, we introduce a basic polling system and discuss the general literature. Subsequently, we present the discrete time polling systems.

3.1. Literature Review on Batch Processes

There is a large body of literature, which studies batch processes. Analysis is conducted mostly in transform domain. In general, we differentiate between batch queues and batch building processes.

The general batch (bulk) queue is described in Bhat (1964) as follows: batches of customers arrive at a single server and get served in batches. The sizes of the incoming batches are independent and identically random variables. The time interval between successive arrivals is assumed to be an iid random variable; so also is the service time of arbitrary batches. The maximum size of the served batch is limited to the capacity of the server. It is commonly assumed that the service time is independent of the queue length at that time. Following Kendall's notation, the notation $G^X/G^Y/1$ is used to represent the general single batch server queue. The exponents X and Y denote the sizes of the arriving batches and served batches respectively. We omit these exponents when they are equal to one. The literature on batch queues is diverse. In order to give a structured overview of the existing literature, we differentiate between:

- Batch arrival queue: batches of customers arrive at the system and get served singly,
- Batch server queue: single customers arrive at the system and get served in batches,
- Batch queue: batches of customers arrive at the system and get served in batches.

For a profound introduction to the batch queuing theory, the works of Chaudhry and Templeton (1983) and Schleyer (2007) are recommended. They analyze, in their books, the systems, in which arrivals, services, or both occur in batches. They discuss and extend existing literature on batch arrival and batch service models and methods. Additionally, application-oriented case studies are included.

Concluding, we will discuss the literature on batch building processes. In the batch building model, incoming customers are collected at a collecting station. Collecting process finishes, when a given rule is satisfied. Then the collected batch departs. At this moment, a new collecting process starts. Unfortunately, the analysis of batch building processes have not attracted much attention from the scientific community, although they are implemented very often in industrial practice.

3.1.1. Batch Arrival Queue

The first work regarding the batch arrival queue is the work of Gaver (1959). He investigates the $M^X/G/1$-queue, in which batches of random size arrive at a single server in exponential inter-arrival times. The service time has an arbitrary distribution. Since then, many authors contributed to the analysis of the $M^X/G/1$-queue in continuous time domain (see Burke (1975) and Van Ommeren (1990)). An extension of the $M^X/G/1$-queue was studied by Madan et al. (2004). They analyze the $M^X/\binom{G_1}{G_2}/1$-queue with batch arrivals, in which the server offers two types of service with different service time distributions. The server serves customers singly and the served customer may ask for a re-service of the service taken by him. They derive the generating functions for the queue length and compute the mean queue length as well as the mean waiting time.

Chaudhry and Gupta (1997) derive the queue length and the waiting time distributions for the discrete $G^X/Geom/1$-queue. Batches arrive at the server in generally distributed inter-arrival times and the service time has a geometric distribution. They consider two possibilities regarding the sequence, in which events are processed. In an early arrival system, a departure is processed before an arrival. The reverse applies in a late arrival system. Lee (2001) considers the discrete time $Geo^X/G/1$-queue with two priority classes, in which service times may differ between the classes. Based on the generating function technique and the supplementary variable method, an analysis of the number of customers in system and the busy period is presented in both sources.

Tran-Gia and Ahmadi (1988) solve the discrete time $G^X/D/1$-queuing system with a finite buffer and a general batch size distribution. The service times are Dirac distributed. They present algorithms to determine the steady state probabilities and the blocking probabilities of the batches and the single units.

For the waiting time of the $GI^X/G/1$-queuing system, Yao et al. (1984) show that the waiting time is the sum of two independent components. One of the components is the same as the waiting time in the $GI/G/1$-system (with single arrivals), so that known results of the $GI/G/1$-queue can be applied readily. Based on this fact, Schleyer and Furmans (2007) present an analytical method to calculate the waiting time distribution for the $G^X/G/1$-queuing system in discrete time domain. Their approach is based on the Wiener-Hopf factorization. Furthermore, they derive insights into the influence of the batch size on the waiting time.

Multi-channel systems with batch arrivals have also been studied by a number of authors. Chaudhry et al. (1992) propose an efficient and reliable method for numerically finding the limiting distribution of the queue length in the $M^X/D/c$-queue. Chaudhry and Kim (2003) present the complete distribution for the system content of a discrete time multi-server queue with batch arrivals and deterministic service times. Furthermore, they derive the waiting time distribution for the case, that the service time equals the unit time. Finally, Choi and Park (1992) and Cong (1994) study the $M^X/M/\infty$ and $M^X/G/\infty$ queues, respectively. Both assume that each batch consists of a stochastic number of customers of k types and there are infinitely many servers in the system. Besides, the service time distributions for different types can be different. Based on generating functions, they compute the number of customers of a specific type served in a fixed time.

3.1.2. Batch Server Queue

The analysis of batch server queues dates back to Bailey (1954). Bailey investigates the $M/G^{[1,K]}/1$-queue. The batch server serves at most K customers simultaneously. He studies the equilibrium distribution of the queue length by means of the embedded Markov chain technique. Additionally, he provides the mean and the variance of the queue length and the mean waiting time. Neuts (1965) studies the busy period of the $M/G^{[1,K]}/1$-queue and shows that the busy period is equal to the time between successive visits to state 0 in a semi-Markov process associated with the queuing process. Based on this, he obtains the distribution of the length of the busy period in transform domain. The waiting time distribution for this queue is represented in Gnedenko and König (1984). Furthermore, Baba (1996) considers the $GI/M^{[1,K]}/1$-queue, in which the service rate depends on the served batch size.

Neuts (1967) introduces the minimum batch size service policy and analyzes the $M/G^{[L,K]}/1$-queue. Complying with the introduced service policy, the server starts the service when there are at least L customers in the queue or waits until L customers are collected. The capacity of the server is limited to K. This means, if the number of waiting customers exceeds K, just K of them receive service immediately. Thus, the queue length is reduced by K customers. He obtains a description of the output process and the distribution of the busy periods. Moreover, he derives the queue length both in discrete and continuous time domains. The results are given in terms of transforms. Gold and Tran-Gia (1993) analyze the $M/G^{[L,K]}/1 - S$, in which the buffer has a limited size S. They stress the manufacturing issues of the finite system $M/G^{[L,K]}/1 - S$, in particular the determination of the threshold value L. They derive the state probabilities and provide insights into dimensioning aspects of production machines. Dümmler (1998) analyzes the departure process for this queue in discrete time domain. He describes the inter-departure and the departing batch size distributions and compute the coefficient of correlation between the inter-departure time and the batch size. It is shown that the inter-departure time distribution of a batch service system has characteristic properties that cannot be approximated by known distributions with sufficient accuracy. This conclusion favors the discrete time analysis, with which it is possible to model arbitrary distributions, rather than an approximation by a theoretical distribution.

Schleyer (2007) drops the Markovian property of the arrival process and derives an approximate method for the discrete $G/G^{[L,K]}/1$-queue based on Dümmler's approach. He derives the inter-departure time, departing batch size, and the waiting time distributions. We will allow in section 4.4 batch arrivals and investigate the $G^X/G^{[L,K]}/1$-queue.

Gupta and Goswami (2002) study the discrete $Geo/G^{[L,K]}/1$-S-queue. This is a finite buffer queue, in which the buffer size is limited to S places. The queue is analyzed and the distribution of the buffer content at departure epochs as well as at arbitrary epochs has been obtained. They consider both early arrival and late arrival policies. Finally, Tran-Gia and Schömig (1996) study a discrete batch server queue under the minimum batch size and the bounded idle time rules. The bounded idle time is commonly used as a job-floor control rule in semiconductor industry. They demonstrate the feasibility of discrete time analysis technique for operational research in manufacturing.

3.1.3. Batch Queue

Bhat (1964) studies the discrete time behavior of the batch queues (i) $M^X/G^Y/1$ (Poisson arrivals and arbitrary service time) and (ii) $GI^X/M^Y/1$ (arbitrary inter-arrival time distribution and negative exponential service time). He obtains the discrete time transition probabilities and the equilibrium behavior of the queue lengths in the systems along with distributions concerning the busy periods. After this pioneering work, batch queues are investigated predominantly in continuous time domain (e.g. Gupta and Goyal (1966), Bagchi and Templeton (1973), Prabhu (1987), Abolnikov et al. (1994)).

Fewer instances, can be found in the literature, in which the batch queues are handled in discrete time domain. Zhao and Campbell (1996) study the discrete $D^X/D^Y/1$ queuing system by means of the generating function technique and obtain the mean number of customers in the system. Alfa and He (2008) study the discrete $GI^X/G^Y/1$ queuing system. Some general results are obtained for the stability condition, stationary distributions of the queue length and the waiting time. For this purpose, algorithmic procedures are developed.

Some works related to the application of the minimum batch size in batch queues appear in the literature. The analysis is mostly conducted under the assumption of multiple vacations. Krishna et al. (1998) derive the system size distribution and the expected lengths of the idle and busy periods of the $M^X/G^{[L,K]}/l$ queuing system with N-policy, multiple vacations and set-up time. The system operates as follows. After finishing a service, if the queue length is less than L, the server leaves for a vacation of random length. After each vacation period, the queue length is checked, if it is at least N ($N \leq K$), then the server starts the service after a set-up time. Otherwise, the server leaves for another vacation. Furthermore, Arumuganathan and Ramaswami (2005) investigate the $M^X/G^{[L,K]}/1$ queuing system with fast and slow service rates and multiple vacations. The service rate is dependent on the queue length. They derive the probability generating function of the queue size at an arbitrary time instant. Various performance measures are obtained. A cost model is discussed and a numerical solution is presented.

There are also models in the literature, related to our analysis regarding the application of batch queues in transport systems. Powell and Humblet (1986) analyze different vehicle dispatching rules by means of batch queuing theory. In many instances, the departures are delayed or canceled in order to avoid low utilization of the vehicle. Specifically, he considers the following rules:

1) No control: vehicles depart regardless of the queue length (the same policy introduced by Bailey (1954))
2) Vehicle cancelation rule: the departure is canceled if the queue length is not sufficient ($< K$). The transport units wait an extra service time.
3) Vehicle holding rule: this rule is the same as the minimum batch size rule introduced by Neuts (1967).
4) Combined rule: if the queue length is not sufficient, the vehicle waits T time units. And if the queue length does not become sufficient within

the time period, the departure is canceled. The vehicle departs as soon as the adequate queue length is achieved without waiting for the end of the time interval.

They present a general theoretical framework for modeling a broad range of vehicle dispatching strategies and derive the transform of the queue length distribution at the end of a service process. Simão and Powell (1988) determine the waiting time distribution under the general vehicle dispatching rule with vehicle holding and vehicle cancelation. They employ discrete time analysis.

Sim and Templeton (1983) consider a transportation network with multiple vehicles, which carry the passengers from a source terminal to different destinations. The system operates under the following policy: when a vehicle is available and there are at least L passengers waiting for service, then a vehicle is dispatched immediately. Assuming that the inter-arrival time of the passengers and the trip-times are exponentially distributed, they attain the queue length and the waiting time distributions. However, they assume that the vehicles have infinite capacities, which is not the case in real applications. Similarly, Lee and Srinivasan (1990) study the control of infinite capacity shuttle for the transport of passengers. The passengers arrive in accordance with a compound Poisson process. When the number of collected passengers exceeds a given limit, the transport process starts. They attain the mean waiting time of an arbitrary passenger. We also investigate a number of vehicle dispatching strategies in section 4.4.5 and compare their performances.

3.1.4. Batch Building Processes

Opposed to the batch queues, batch building processes have attracted little attention from the scientific community. Having employed the decomposition approach, Bitran and Tirupati (1989) investigate the $G/G^{[K,K]}/1$-queue with multiple product classes. Based on this approach, the system was decomposed into a batch building node, in which batches of fixed size (K) are collected, and a server node. They investigate the departure process at the batch building node in continuous time domain and attain the first two moments of the inter-departure time. Similarly, Fowler et al. (2002) study a batch building node also owing to the decomposition approach. With the aim of analyzing a multiproduct batch service queue, they compute the mean waiting time of a batch building node, in which items were grouped into batches of product dependent fixed sizes. Apart from these authors, who model batch building operations as an intermediate result of the decomposition approach, Meng and Heragu (2004) investigate a batch building operation of fixed size in continuous time domain.

To our very best knowledge, Schleyer (2007) is the first to analyze batch building processes in discrete time domain. He studies two basic batch building modes, the capacity and the timeout rule, and additionally a modification of these basic modes, called the minimum batch size rule. Under the capacity rule, customers are collected until a given collection quantity is obtained. On the other hand, collecting time is the deciding criterion under the timeout rule. This means, that customers are collected until a given length of time elapses. Finally, minimum batch size rule is defined as follows. The collecting process lasts at least a given length of time. When this time interval ends and the number of collected customers is less than a given minimum quantity, then the collecting process continues till the minimum quantity is reached. In sections 4.2 and 4.3, new batch building rules are introduced.

3.2. Literature Review on Polling Systems

A polling system is characterized by multiple queues, attended by a single server. In its most native form, the server visits the queues in a cyclic order to render service to the waiting customers at different queues. Upon the completion of service at a given queue, the server incurs typically a certain amount of time to move to the next queue, which is referred to as the switch-over time.

Polling systems find applications in diverse fields. This is not a coincidence, as in many applications the users compete for a common resource (server) and the cyclic service yields a fair allocation of the resource among the users. The history of the polling models dates back to the late 1950s. The first publications from Mack et al. (1957) and Mack (1957) involve the analysis of the patrolling repairman model for the British cotton industry. Since then, numerous polling systems have been studied in the literature and used for the performance analysis

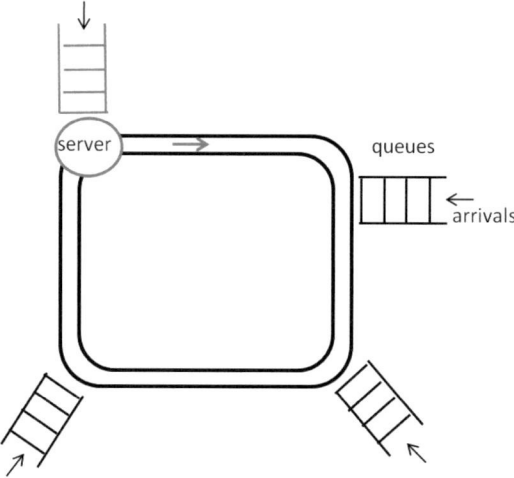

Figure 3.1.: A basic polling system

and the optimization of a wide variety of applications in computer, telecommunication, production, and transportation systems. According to Takagi (2000), the field of polling systems proved to be one of the few successful theoretical performance evaluation models developed in the last decades.

Similar to the discussion in Takagi (2000), we present here a basic polling model in discrete time domain. Figure 3.1 displays a basic polling system. The system, we consider, consists of N queues with identical characteristics. The customers arrive at each queue in stochastic inter-arrival times. The discrete RV A denotes the inter-arrival time. We assume that the service time of a customer at each queue is described by the discrete RV B. Both variables are iid. The basic polling system is a symmetric system, which means that the queues have exactly the same characteristics. In this system, the server serves the queues in a cyclic manner in the order $Q_0, Q_1, \cdots, Q_{N-1}, Q_0, Q_1, \cdots, Q_{N-1}, \cdots$ It is assumed that the switch-over time from an arbitrary queue i to the next queue is denoted by the discrete RV S. For this system, the arrival rate (λ) is computed as the reciprocal of the mean inter-arrival time

($\lambda = 1/E(A)$). Analogous to that, the service rate is the reciprocal of the means service time ($\mu = 1/E(B)$). Based on these values, the total utilization of the system is given by ($\rho = N\lambda B$). Another important performance measure in such a system is the cycle time (C), that the server needs to complete a cycle by serving all the queues in the system. The cycle time corresponds also to the time interval between successive service processes at an arbitrary queue. Moreover, the waiting time (W) of an arbitrary customer from its arrival to service start is relevant for the performance analysis of the polling systems.

Due to the wide-spread application of polling models in a variety of settings in different fields, many variants of the basic models have been studied. In order to give a structural overview of the existing polling models, we consider the following characteristics:

- Buffer size: Single/Finite/Infinite buffer systems
- Service discipline: Exhaustive/(Globally) Gated/Limited/Decrementing
- Symmetric/Asymmetric systems
- Server routing:
Cyclic/Random/Periodic/Dynamic/Priority order etc.
- Queuing discipline: Non-Priority/Priority queues
- Number of servers: Single/Multiple server(s)
- Networks
- Special application area: Production/Traffic/Transport systems etc.

We address now these characteristics and possible variants briefly. However, emphasis is given to the models, which are relevant to the current work regarding either the employed analysis approach or the application related aspects. For an extensive discussion of polling systems, reviews can be found in Takagi (1986), Takagi (1988), Takagi (2000), Levy and Sidi (1990), Adan et al. (2001), and Vishnevskii and Semenova (2006).

Buffer size

There is a huge literature on polling models, in which queues are assumed to have infinite buffer sizes. In some others, the buffers have finite sizes (see Tran-Gia and Raith (1988), Takagi (1991)). In such systems, the customers, that find the buffer full, get lost. A special case of finite buffer systems is the single buffer systems, in which queues have unit-sized buffers. Single-buffer systems are used commonly in repair problems, in which the queue length becomes one, if a machine fails.

Service discipline

The service discipline specifies the rule, according to which the server quits the service and leaves an arbitrary queue. The variants can be summarized as follows:

- **Exhaustive service discipline**: complying with this discipline, the server does not leave the queue until it becomes empty.

- **Gated service discipline**: under the gated service regime, the server treats only the customers that were already present in the queue at the polling instant. The customers that arrive at the queue after the start of the service are left for the next cycle. An extension of the gated service is the globally gated regime. Under this regime, the server uses the beginning of the cycle as the reference point and serves at each queue only those customers that were present in the queue at the reference point (see Boxma et al. (1992)). Levy and Sidi (1991) consider polling systems, at which customers arrive in batches. Under the assumption of general service and switch-over times, they derive, for both the exhaustive and the gated service disciplines, the mean waiting time. For a discussion of exhaustive and gated service disciplines, see also Cooper and Murray (1969), Eisenberg (1972), Hashida (1972), Ferguson and Aminetzah (1985), Srinivasan et al. (1996), Winands et al. (2006).

- **Limited service discipline**: there are two variants of the limited service discipline. Under the K-limited service rule, the server can serve at most K customers at an arbitrary queue. The second variant is the time-limited service rule, in which the duration of the server attendance to each queue is limited. The mathematical treatment of the limited service disciplines is more difficult than the gated and the exhaustive service regimes (Takagi (1988)). Limited service disciplines are discussed in Fuhrmann (1985), Leung (1991), Lee and Sengupta (1992), and Leung (1994).

- **Decrementing service discipline**: under the decrementing (semi-exhaustive) service discipline, service starts if the queue is not empty and continues until the queue length is decreased by one customer.

Besides above mentioned service disciplines, numerous hybrid regimes are defined and analyzed in the literature. These hybrid disciplines are a mixture of the conventional service disciplines, e.g. gated limited service discipline (see Dittmann and Hübner (1993)). For a broad class of polling systems including the conventional disciplines and their hybrids, Kuehn (1979) shows that the mean cycle is computed as follows

$$E(C) = \frac{E(S^*)}{1 - \rho}$$

where S^* is the total switch-over time in a cycle.

Symmetric/Asymmetric systems
In the asymmetric systems, each queue has different characteristics regarding e.g. the inter-arrival time, buffer sizes etc. On the other hand, the queues and their treatment in symmetrical systems are identical. Takagi (1988) demonstrates the effects of different service regimes on the mean waiting time both in symmetric and asymmetric systems. In accordance with that, mean waiting times in symmetric systems under different regimes have the following relation to each other:

$$E(W)_{\text{exhaustive}} \leq E(W)_{\text{gated}} \leq E(W)_{\text{limited}}$$

$$E(W)_{\text{exhaustive}} \leq E(W)_{\text{decrementing}} \leq E(W)_{\text{limited}}$$

In asymmetric systems, heavily loaded queues experience lower waiting times under the exhaustive regime, whereas the gated and the limited service strategies prevent this effect.

Server routing
The regime of the server routing defines the order, in which the queues are served. Under cyclic routing, the order of service is deterministic. However, there are other systems, in which the order changes. An example of such a regime is the random polling, in which the next queue to be visited is queue j with a probability p^j (see Kleinrock and Levy

(1988)). A generalization of the random polling is the Markovian routing scheme, in which the server switches from queue i to queue j with a given probability $p^{i,j}$ (see Boxma and Weststrate (1989) as well as Srinivasan et al. (1996)).

Another possibility is that the server visits the queue according to a given polling order table. This kind of polling is also called the periodic polling as the polling order table, which has a finite length, is repeated (see Baker and Rubin (1987), Olsen and Van der Mei (2003)).

Under the dynamic server routing, the decision on the next queue is made based on e.g. the value of a variable. For instance, the next queue to serve can be chosen dependent on the queue length. Finally, in the priority order routing, queues have different priorities. A queue can only be served when the buffers of the queues with higher priorities are empty (see Lye and Seah (1992) and Chakravarthy (1998)).

Queuing discipline
The customers at the same queue may have different priorities regarding the order of service (see Boon et al. (2008) and Boon and Adan (2009)). Moreover, different service disciplines can be applied for different queues in asymmetric systems. For example, it is possible to apply exhaustive service discipline for some queues and apply the gated strategy for the others. In this way, the queues, at which exhaustive discipline is applied, can be prioritized (see Ferguson and Aminetzah (1985)).

Number of servers
Besides the traditional systems with a single server, systems with multiple servers exist (see Browne and Weiss (1992) as well as Van der Mei and Borst (1997)). These servers can be identical or non-identical. Such models are applicable to transport systems.

Networks
In the literature, many publications can be found, in which different network assumptions are made. For illustration, in closed polling models, it is assumed that a constant number of customers circulates in the system (see Altman and Yechiali (1994)).

Special application area

Although numerous applications of polling models can be found in a variety of systems, we concentrate here only on the production, traffic, and transport systems. A special emphasis is given to the applications in transport systems.

Production systems: production systems, in which multiple products with random demands are served by a single machine, can be modeled as polling systems. The production orders correspond in such a polling system to the customers and the different products to the queues. The demand distributions make up the arrival process at each queue. In make-to-stock production systems, each product is kept in stock. In many real production systems, a base-stock level is assigned to each product, which is the targeted inventory level for the product. As a result, each product is served until the given base-stock level is attained. For a discussion of such models, see Federgruen and Katalan (1996a), Federgruen and Katalan (1996b), Federgruen and Katalan (1998), Krieg and Kuhn (2002), and Grasman et al. (2008). In contradiction to the make-to-stock systems, products are not stocked in make-to-order environments, simplifying the modeling effort. Recently, Boon and Adan (2009) introduce a mixed exhaustive-gated strategy. Complying with this strategy, two priority classes are defined. In particular, high priority customers are served complying with the exhaustive discipline, whereas low priority ones are served under the gated discipline. Furthermore, high priority customers are always served ahead of the low priority ones. This service strategy can be used to model make-to-order production systems, in which products for both internal and external customers are produced according to a fixed production sequence and external customers have the priority over the internal customers.

Traffic systems: polling systems find applications for the analysis of traffic systems, frequently when a traffic intersection is used by multiple flows. In such a system, each direction is viewed as a queue. When the traffic light becomes green for a traffic direction, the cars (customers) pass the traffic light. The driving time to pass the traffic light corresponds here to the service time. The subset of publications with application to traffic systems includes Newell (1998) and Lehoczky (1972).

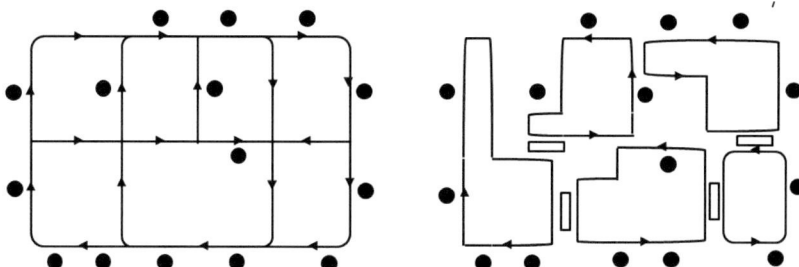

Figure 3.2.: Left: example of a conventional AGV system; right: example of a tandem AGV system. A tandem AGV system is obtained by partitioning the pick-up/deposit stations into non-overlapping single-vehicle loops

Transport systems: the majority of the literature on transport system involves the analysis of automated guided vehicles (AGV). Automated guided vehicles are used typically in material handling systems to move raw materials or finished products. In the industry, it is possible to encounter AGV systems with varying level of complexity. In some systems, there is only one vehicle that serves a couple of work stations, whereas some others involve a complex network of work stations, a number of vehicles, and a control system to assign the vehicles to the stations. Consequently, many design questions arise regarding e.g. the vehicle routing, which are quite often handled by means of polling models.

In an AGV system, the vehicle is the server, and the stations are the queues. As mentioned above, there are usually multiple vehicles in a system. Dukhovnyy (1979) considers such a transportation system and models it as a system with a single vehicle, which moves at a faster rate.

The routing of the AGV plays an important role in the performance of the system. The routing can be centralized on a first come first served (FCFS) basis, in which the vehicle moves to the queue with the oldest waiting customer. Although it is perceived as fair, this strategy is inefficient due to resulting empty travels. On the other hand, Bartholdi and Platzman (1989) introduces a decentralized decision rule, first encountered first served rule (FEFS). According to this rule, the vehicle

27

circulates a loop until it finds the first queue to be served. Bozer and Srinivasan (1991) decompose a multi-vehicle network into single-vehicle closed loops that operate under FEFS (see figure 3.2). That's why no control system is needed to control these single loops. This configuration is called the tandem configuration. For the tandem configuration, they discuss the throughput performance of a single loop. Bozer et al. (1994) derive an approximate expression for the mean waiting time for pick-up at an arbitrary station in a single loop.

Having discussed the general literature on polling systems, we concentrate in the next section on the discrete time polling systems.

3.2.1. Discrete Time Polling Systems

Although the bulk of the literature on polling systems is devoted to continuous time systems, many polling systems have also been studied in discrete time domain. One of the earlier contributions in discrete time domain is the work of Konheim and Meister (1974), which contributes an exact analysis of exhaustive service systems. They assume symmetric queues and nonzero switch-over times. Levy and Kleinrock (1991) analyze a polling model also under exhaustive service regime with zero switch-over times.

Some authors study limited service disciplines under various assumptions. Tran-Gia (2002) develops an approximate algorithm for polling systems under the assumptions of non-exhaustive service (or more precisely limited service of one customer per service process), general renewal input traffic, and finite capacity waiting places. The analysis is based on the discrete time convolution operations using Fast Fourier Transform (FFT). Takagi and Leung (1994) study a single server vacation system, in which the server renders time-limited service and takes a vacation when the limit expires or the queue empties. They study the process of the unfinished work and the joint process of the queue size and the remaining service time and obtain the mean waiting time.

Dittmann and Hübner (1993) analyze a cyclic server system under the gated limited service discipline. They assume general renewal input traffic, service time, and switch-over time, and present an approximate iterative algorithm to compute the cycle time and queue state distributions. Moreover, they derive the waiting time distribution for an arbitrary customer approximately. Fiems et al. (2002) investigate a vacation system under the gated-exhaustive strategy. They obtain expressions for the moments of the system contents at various epochs and for the customer delay.

Server routing is also considered in discrete polling systems. Kleinrock and Levy (1988) analyze the behavior of random polling systems, in which the next station to be served after station i is determined by probabilistic means. Specifically, the random polling scheme is studied for three types of service policy: 1) exhaustive service, 2) gated service, and 3) limited service. They derive expressions for the expected response time in a random polling system operating under these service disciplines. Boxma et al. (1990) investigate a single-server polling system, in which the stations are polled according to a general service-order table. They consider the exhaustive, gated, and 1-limited strategies. Stability conditions were established and an expression for the mean waiting times is attained. An analysis, in which batch arrivals are considered is presented by Beekhuizen and Resing (2009). The server follows the Markovian routing and the Bernoulli service regime. Bernoulli service means that after the service of a customer at an arbitrary queue i, the server serves the same queue again with probability $q(i)$ and moves to another queue with probability $(1 - q(i))$. For this system, they derive marginal queue length distribution approximately, based on the translation of the polling system to a structured Markov chain.

Finally, Takahashi and Kumar (1995) study a priority polling system, in which customers of different priority classes arrive at each station according to independent Bernoulli batch processes. The head-of-the-line (HL) priority rule and nonzero switch-over times between stations are assumed. The customers are served at each station under a mixed (exhaustive, gated, and 1-limited) service strategy. They study the mean waiting times for the priority classes.

.

4. Inventory Consolidation

4.1. Introduction

In transport systems, economies of scale can be realized as unit transport costs decrease with increasing shipment size. Therefore, small shipments are often consolidated into large ones and transported in large quantities. This form of consolidation is known as the inventory consolidation, in which shipments are delayed until a predefined state is reached. Specifically, shipments are accumulated until a specific volume is reached or a given time interval has elapsed. In this way, large transport quantities are attained. However, the effect of additional delay time should be studied. In particular, it should be validated, that the cost reductions achieved by the consolidation suffices to cover the increased inventory costs or the service deterioration caused by the delay time. Hence, efficient methods are needed to quantify the effects of inventory consolidation on the service level or the total costs.

In this chapter, we represent models for the analysis of inventory consolidation. The chapter is structured as follows: in sections 4.2 and 4.3 we represent exact models for the batch building modes, the capacitated timeout rule and the capacity interval rule. We derive the waiting time, inter-departure, and the departing batch size distributions for both rules. In section 4.4, we study a batch queue, which is denoted as the $G^X/G^{[L,K]}/1$-queue in Kendall's notation. For the $G^X/G^{[L,K]}/1$-queue, we derive the inter-departure time, departing batch size and the waiting time distributions. The introduced methods are exact within an ϵ-environment. Eventually, we show how these models can be applied for the analysis of different inventory consolidation strategies.

4.2. Batch Building: Capacitated Timeout Rule

This section is devoted to the analysis of the batch building rule "Capacitated Timeout Rule". As the name implies, it is an extension of the timeout rule (Schleyer 2007) where a limiting capacity applies. Therefore, the collecting process ends when either the maximum collecting size or the maximum collecting time is reached. As the focus of our analysis is on the transport systems, we refer to the incoming objects as the transport units for the rest of our analysis. However, the model is applicable to a broad range of systems.

In this model, we assume that the transport units arrive at the collecting station in batches of stochastic size and in stochastic time intervals. The incoming transport units are collected at the collecting station till the end of the collecting process. Then, the collected batch is transported to the next station and a new collecting process starts immediately at the collecting station. This kind of batch building mode is widespread in transport systems as the capacities of the transport vehicles are limited and usually a maximum collecting time is defined to limit the waiting time of the transport units. Upon the departure of one vehicle, a new vehicle becomes available and a new collecting process starts.

4.2.1. Queuing System

We consider the batch building process under the capacitated timeout rule in discrete time domain. Figure 4.1 illustrates the collecting process and some variables. We assume that the arrival process is defined by two iid discrete variables; inter-arrival time (A) and incoming batch size (Y). Furthermore, we denote the maximum collecting time and the maximum collecting size as t_{out} and K, respectively. For this mode, we derive the inter-departure time, and the departing batch size distributions. These distributions are needed for the network analysis as they describe the arrival process at the succeeding queue. Moreover, we derive the waiting time distribution, which can be used to derive the sojourn time distribution or to assess the inventory costs. We summarize below the variables and the parameters used in this analysis.

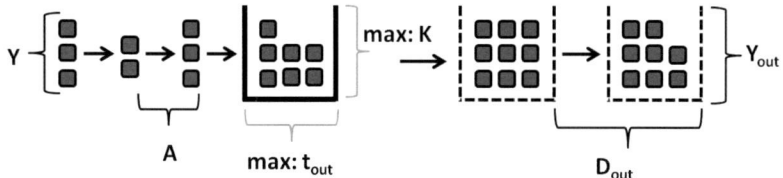

Figure 4.1.: Collecting process under the capacitated timeout rule

t_{out} maximum collecting time,
K maximum collecting size,
A inter-arrival time, time between two successive arrivals,
Y batch size of an incoming batch,
R^y remainder, number of transport units left over from the previous collecting process due to the limited collecting size,
R^a residual inter-arrival time, time interval between the start of a collecting process and the first arrival within this process,
R residual state, two dimensional variable defining the states of residual inter-arrival time and remainder at the beginning of a collecting process,
D_{out} inter-departure time between two successive departures,
Y_{out} departing batch size,
W waiting time of an arbitrary transport unit.

It is assumed that $a_{max} \leq t_{out}$ and $y_{max} \leq K$. The first assumption assures that at least one arrival occurs in each collecting period. On the other hand, as $y_{max} \leq K$, at least one arrival is needed to complete a collecting process. In other words, the number of remaining customers is never sufficient to fill up the capacity available and finish the collecting process in this way. Besides, zero-fold convolutions of the input distributions are assumed to be Dirac distributed with a constant value of zero, thus, $y_0^{0\otimes} = a_0^{0\otimes} = 1$.

Since the process development is observed in the discrete time domain, several events may occur simultaneously. Consequently, the events are processed in the following sequence: 1) arrival 2) end of a collecting process 3) start of a collecting process.

Residual inter-arrival time and remainder distributions

Residual inter-arrival time (R^a) is defined as the time interval between the start of a collecting process and the first arrival in this collecting process. It can be either a complete inter-arrival time or a residual of it. Similarly, remainder (R^y) stands for the number of transport units, which were left over from the previous collecting process and are, therefore, already present in the queue at the beginning of the subsequent collecting process.

In order to derive the distributions of the residual inter-arrival time and the remainder, two main cases of the collection process should be investigated. In the first case, less than K transport units are collected within the maximum collecting time t_{out} in an arbitrary n^{th} collecting process. In this case, the collecting process finishes in exactly t_{out} time units and all the collected units depart together as a batch. Consequently, the number of remaining units at the beginning of the $(n+1)^{th}$ process is zero and the residual time is either a whole inter-arrival time (only if the last batch arrives exactly at the end of t_{out}) or a residual of it. Thus, the maximum value of the residual inter-arrival time corresponds to the maximum inter-arrival time (a_{max}). In the second case, the collecting process finishes in $m \leq t_{out}$ time units as the maximum collecting size is exhausted. In this case, the last batch sees $(K - i)$ transport units at the collecting station at the arrival instant. If it has a batch size of $(i+j)$, R^y equals j. In other words, j transport units are left for the $(n+1)^{th}$ collecting process. Note that the maximum value of the remainder is restricted to $(y_{max} - 1)$ as the number of missing units immediately before the arrival of the last batch must be at least one. Given that the last batch completes the number of collected transport units to K, then the remainder equals zero. Under the second case, the collecting process finishes always with a batch arrival. That's why, the next residual inter-arrival time corresponds to a complete inter-arrival time. Table 4.1 summarizes the possible values for the residual inter-arrival time and the remainder under each case.

The distributions of the residual time and the remainder must be known to derive further performance measures e.g. inter-departure time distribution. The major challenge arises from the fact that the remainder

Case	Number of collected units	Time elapsed from the beginning of the process	Remainder	Residual time
1	$< K$	t_{out}	0	$[1, a_{max}]$
2	$\geq K$	$m \ (m \leq t_{out})$	$[0, y_{max} - 1]$	$[1, a_{max}]$

Table 4.1.: Possible courses of the collecting process under the capacitated timeout rule

at the beginning of an arbitrary $(n + 1)^{th}$ process is dependent on the residual inter-arrival time and the remainder of the nth process. Likewise, the residual inter-arrival time of the $(n + 1)^{th}$ process depends also on the residual inter-arrival time and the remainder of the nth process. These dependencies make it complicated to compute the remainder and the residual inter-arrival time distributions separately. A way to circumvent this problem is to replace the remainder and the residual inter-arrival time distributions with the residual lifetime of a discrete renewal process. Subsequently, the inter-departure time, departing batch and the waiting time distributions can be estimated only approximately. Instead, we present an exact approach.

For the remainder of our analysis, we define a new variable and refer to as residual state (R). As the name implies, the residual state is a two-dimensional variable defining the states of the residual inter-arrival time and the remainder at the beginning of an arbitrary process. Correspondingly, the joint probability $r_{s,j}^n$ describes the probability that the residual inter-arrival time assumes the value s and the remainder j simultaneously at the beginning of the nth process. This probability depends noticeably on the residual state from the $(n - 1)^{th}$ process but not on that of the $(n - 2)^{th}$ process. Consequently, this process corresponds to a Markov process and can be analyzed by means of a bivariate (discrete) Markov chain. In figure 4.2, the Markov chain for a case with $a_{max} = 2$ and $y_{max} = 2$ is given.

The state space of such a Markov Chain is finite and limited to $a_{max} \cdot y_{max}$ states (for the given example $a_{max} \cdot y_{max} = 4$). Moreover, every state can be reached from another directly or indirectly. Such Markov chains are referred to as irreducible Markov chains. As stated by Gnedenko and König (1983), an irreducible and aperiodic Markov

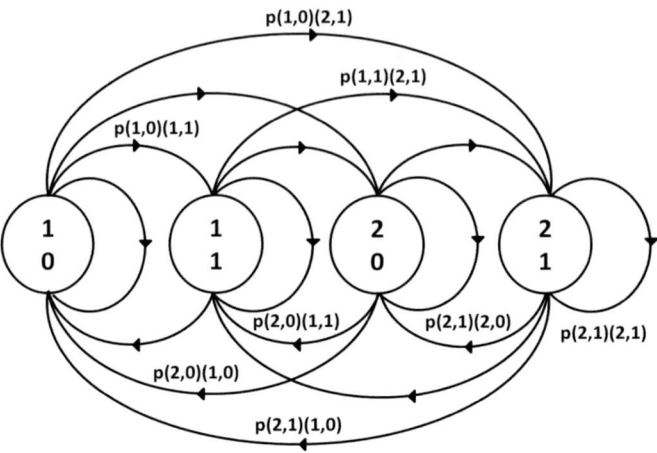

Figure 4.2.: Discrete Markov chain with possible states
for $a_{max} = 2$ and $y_{max} = 2$ for the analysis of the residual state under
the capacitated timeout rule

chain with limited number of states is ergodic. For an ergodic Markov chain, the steady state distribution exists and can be assessed with the aid of stationary equations

$$r_{s,j} = \sum_{u=1}^{a_{max}} \sum_{i=0}^{y_{max}-1} p(u,i)(s,j) \cdot r_{u,i} \tag{4.1}$$

where $p(u,i)(s,j)$ stands for the transition probability from residual state (u,i) to (s,j). Moreover, the sum of probabilities is equal to 1.

$$\sum_{s=1}^{a_{max}} \sum_{j=0}^{y_{max}-1} r_{s,j} = 1 \tag{4.2}$$

With equations 4.1 and 4.2, an overdetermined equation system is obtained. Hence, one of the equations has to be omitted. Before the equation system can be solved, the transition probabilities must be known. The transition probabilities are computed as follows

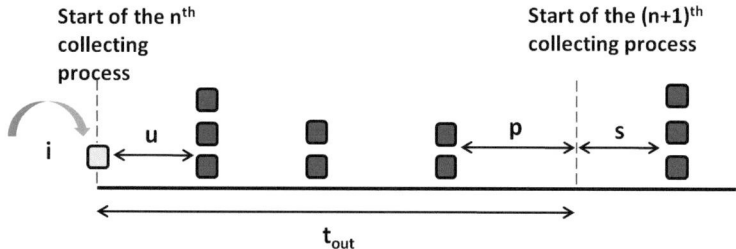

Figure 4.3.: Example of a collecting process under the capacitated timeout rule, which ends in exactly t_{out} time units with number of collected units $< K$

for $j = 0$,

$$
\begin{aligned}
p(u,i)(s,j) \quad &= \sum_{l=1}^{l_{max}} \sum_{p=0}^{t_{out}-u} \sum_{x=1}^{K-i-1} a_{t_{out}-u-z}^{(l-1)\otimes} \cdot a_{z+s} \cdot y_x^{l\otimes} \\
&+ \sum_{m=u}^{t_{out}} \sum_{l=1}^{l_{m,max}} \sum_{n=1}^{K-i} a_{m-u}^{(l-1)\otimes} \cdot a_s \cdot y_{K-i-n}^{(l-1)\otimes} \cdot y_n \quad (4.3)
\end{aligned}
$$

In equation 4.3, we consider the collecting process, at the end of which there are no transport units left over ($j = 0$). In the first part of the equation, we presume that l arrivals occur in exactly t_{out} time units. In this case, the number of collected transport units after l arrivals is less than K. Such a process is illustrated in figure 4.3. Assuming a K value of 10 for the case displayed in the figure, there will no units left for the next collecting process ($j = 0$). The variable i stands for the remainder at the beginning of the process. Similarly, u is the residual inter-arrival time. This kind of collecting process does not have to finish with a batch arrival. The variable z defines here the time between the arrival of the last batch and the end of the maximum collecting time t_{out}. If the time interval between the last arrival in the current collecting process and the first arrival in the next process is equal to $(z + s)$, then the resulting residual time equals s. The residual inter-arrival time is a complete inter-arrival time only if the last batch arrives exactly at the end of (t_{out}), yielding $p = 0$. Otherwise the residual inter-arrival time s

is a residual of the inter-arrival time. In the second part of the equation, we assume that l arrivals occur in $m \leq t_{out}$ time units and the last batch completes the number of collected units to exactly K. Hence, the batch building process finishes exactly at the arrival instant of the last batch. Such a process is illustrated in figure 4.4. The number of collected units according to the figure is equal to $K = 10$. So all units depart together.

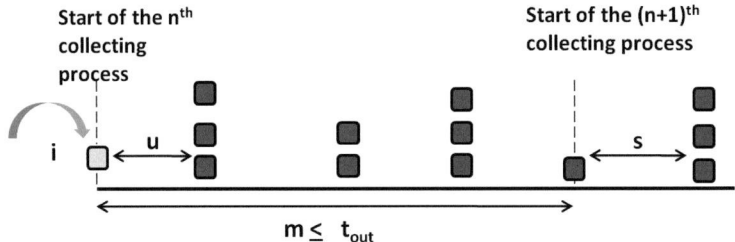

Figure 4.4.: Example of a collecting process under the capacitated timeout rule, which ends in $m \leq t_{out}$ time units with number of collected units $= K$

Note that $l_{m,max}$ and l_{max} stand for the maximum number arrivals that can be observed in each case. $l_{m,max}$ is the maximum number of arrivals given that the collecting process takes m time units, whereas l_{max} is the maximum number of arrivals in the case of a collecting process that continues for t_{out} time units. l_{max} and $l_{m,max}$ are computed by means of the following formulas which apply for the rest of our analysis.

$$l_{m,max} = \lceil \frac{m}{a_{min}} \rceil \qquad l_{max} = \lceil \frac{t_{out}}{a_{min}} \rceil \qquad (4.4)$$

Now, we investigate the case, in which the resulting remainder has a positive value.

for $j \neq 0$,

$$p(u,i)(s,j) = \sum_{m=u}^{t_{out}} \sum_{l=1}^{l_{m,max}} \sum_{n=1}^{K-i} a_{m-u}^{(l-1)\otimes} \cdot a_s \cdot y_{K-i-n}^{(l-1)\otimes} \cdot y_{n+j} \qquad (4.5)$$

In equation 4.5, it is assumed, that l arrivals take place in $m \leq t_{out}$ time units. Here the number of collected transport units after $(l-1)$ arrivals is less than K and the missing number of units is given by n. The last batch has a batch size of $(n+j)$ and completes the number to $(K+j)$ yielding a remainder of j units. This kind of collecting process has, indeed, the same course of development like the second part of equation 4.3. The only difference is that some units are left for the subsequent process. An illustration of such a process is displayed in figure 4.5. If $K = 10$, the remainder becomes one.

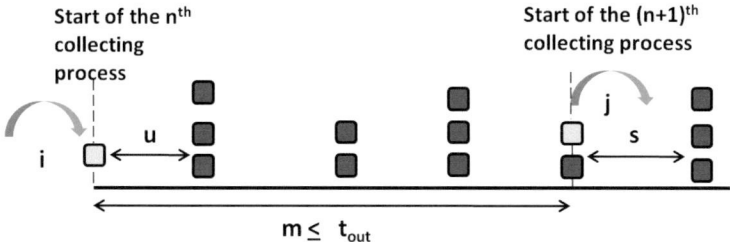

Figure 4.5.: Example of a collecting process under the capacitated timeout rule, which ends in $m \leq t_{out}$ time units with number of collected units $> K$

Solving the resulting equation system in equations 4.1 and 4.2, the residual state distribution is determined.

4.2.2. Inter-departure Time Distribution

Under the capacitated timeout rule, inter-departure time is the collecting time, of which the maximum value is limited to t_{out}. To begin with, we derive the conditional probability that the inter-departure time is equal to the residual inter-arrival time, given the residual state (u, i). In this case, the collecting process finishes with the arrival of the first batch. As the remainder equals i, this batch must have a batch size of at least $(K - i)$. Hence,

$$P(D_{out} = u \mid R = (u, i)) = \overline{y_{K-i}} \tag{4.6}$$

As the next step, we investigate the conditional probability that the inter-departure time has an arbitrary value m $(u < m < t_{out})$ under the condition of a given residual state (u, i).

$$P(D_{out} = m \mid R = (u, i)) = \sum_{l=2}^{l_{m,max}} \sum_{j=1}^{K-i-1} a_{m-u}^{(l-1)\otimes} \cdot y_{K-i-j}^{(l-1)\otimes} \cdot \overline{y_j} \qquad (4.7)$$

Thereafter, we derive the conditional probability for the case that the inter-departure time equals t_{out} time units. The first term in the following expression depicts the case, in which less than K transport units are collected within t_{out} time units. Conversely, the last term represents the case, in which the size of the departing batch is completed to K or more exactly at the end of t_{out} time units.

$$P(D_{out} = t_{out} \mid R=(u,i)) = \sum_{l=1}^{l_{max}} \sum_{p=0}^{t_{out}-u} \sum_{x=1}^{K-i-1} a_{t_{out}-u-p}^{(l-1)\otimes} \cdot \overline{a_{p+1}} \cdot y_x^{l\otimes}$$

$$+ \sum_{l=2}^{l_{max}} \sum_{j=1}^{K-i-1} a_{t_{out}-u}^{(l-1)\otimes} \cdot y_{K-i-j}^{(l-1)\otimes} \cdot \overline{y_j} \qquad (4.8)$$

Finally, with the law of total probability, inter-departure time distribution is computed:

$$P(D_{out} = s) = \sum_{u=1}^{a_{max}} \sum_{i=0}^{y_{max}-1} P(D_{out} = s \mid R = (u, i)) \cdot r_{u,i} \qquad (4.9)$$

4.2.3. Departing Batch Size Distribution

The departing batch size is bounded to K transport units. Under the assumption of a specified residual state, following conditional probabilities are obtained.

$$P(Y_{out} = K \mid R = (u, i)) = \sum_{m=u}^{t_{out}} \sum_{l=1}^{l_{m,max}} \sum_{j=1}^{K-i} a_{m-u}^{(l-1)\otimes} \cdot y_{K-i-j}^{(l-1)\otimes} \cdot \overline{y_j} \qquad (4.10)$$

$$P(Y_{out} = K - j \mid R = (u, i)) = \sum_{l=1}^{l_{max}} \sum_{p=0}^{t_{out}-s} a_{t_{out}-s-p}^{(l-1)\otimes} \cdot \overline{a_{p+1}} \cdot y_{K-i-j}^{l\otimes}$$

$$(4.11)$$

Eventually, we apply the law of total probability. It yields:

$$P(Y_{out} = m) = \sum_{u=1}^{a_{max}} \sum_{i=0}^{y_{max}-1} P(Y_{out} = m \mid R = (u,i)) \cdot r_{u,i} \qquad (4.12)$$

4.2.4. Waiting Time Distribution

In this section, we derive the waiting time distribution of an arbitrary transport unit, of which the maximum value is equal to t_{out}. We follow a similar approach, that is developed by Schleyer (2007) for the derivation of the waiting time under the minimum batch size rule. As the first step, we calculate the probability that the transport unit departs as a member of a collected batch with a size $< K$ transport units $(P(\widetilde{Y_{out}} < K))$. Note that this probability is different than the probability that a collected batch size $< K$ transport units is observed $(P(Y_{out} < K))$. The difference is that the probability $(P(\widetilde{Y_{out}} < K))$ is observed from the view of an arbitrary transport unit and increases with the increasing departing batch size. In other words, it is more probable that an arbitrary unit departs as a part of a larger batch. Thus, this probability is given by

$$
P(\widetilde{Y_{out}} < K) = \sum_{s=1}^{a_{max}} \sum_{i=0}^{y_{max}-1} \sum_{l=1}^{l_{max}} \sum_{p=0}^{t_{out}-s} \sum_{j=1}^{K-i-1} r_{s,i} \cdot
$$
$$
a_{t_{out}-s-p}^{(l-1)\otimes} \cdot \overline{a_{p+1}} \cdot y_{K-i-j}^{l\otimes} \cdot \frac{K-j}{E(Y_{out})} \qquad (4.13)
$$

Comparatively, the probability that an arbitrary customer departs in a batch of K units, is determined as follows

$$
P(\widetilde{Y_{out}} = K) = \sum_{s=1}^{a_{max}} \sum_{i=0}^{y_{max}-1} \sum_{m=s}^{t_{out}} \sum_{l=1}^{l_{m,max}} \sum_{j=1}^{K-i} \sum_{n=0}^{y_{max}-j} r_{s,i} \cdot
$$
$$
a_{m-s}^{(l-1)\otimes} \cdot y_{K-i-j}^{(l-1)\otimes} \cdot y_{j+n} \cdot \frac{K}{E(Y_{out})} \qquad (4.14)
$$

41

It is noteworthy, that both equations are weighted with the relevant departing batch sizes; $(K - j)$ and K. As mentioned above, the probability of belonging to a departing batch increases with its size. Besides, both equations are divided by the mean value of the departing batch $(E(Y_{out}))$. In this way, probabilities are normalized. Afterwards, we proceed as follows. We consider the waiting time for each arbitrary k^{th} batch. We weight the expressions for the waiting time with the number of units that belong to the kth batch and depart at the end of the process in the subsequent departing batch (recall that we investigate the case that the arbitrary unit departs at the end of the collecting process, i.e. does not remain at the collecting station). This number is equal to the size of kth incoming batch for all the batches prior to the last batch. For the last batch, this number may be different than the batch size as some units from this batch can be left over after the departure. In this way, we attain the proportional expressions for the waiting times of the arbitrary batches. Eventually, we normalize these proportional expressions to attain the exact waiting time distribution. This procedure is similar to computing the joint probability that X units from the kth batch wait for Y time units. We multiply this joint probability with X and normalize the expressions.

Now, we consider the waiting time of an arbitrary transport unit under the condition that this unit departs in a batch with a size $< K$ transport units. Subsequently, we assume that the process starts with a remainder of i transport units and the first batch arrives in s time units after the start of the collecting process. Furthermore, it is assumed that l arrivals take place within t_{out} time units and the arbitrary unit belongs to the kth batch (where $k = 0$ stands for the remainder). Since this kind of batch process does not have to finish with a batch arrival, the variable p is defined as the time between the last arrival and the end of t_{out} time units. Figure 4.6 visualizes the process.

Now, we investigate the waiting time for the remaining customers, As we mentioned beforehand, if there are remaining transport units at the beginning of a collecting process, this means that the previous collecting process finished with arrival of the last batch. Consequently, the remaining units arrived at the station exactly at the beginning of the current collecting process. Thus, the waiting time of the remaining customers is t_{out} time units. The proportional value for this probability is computed

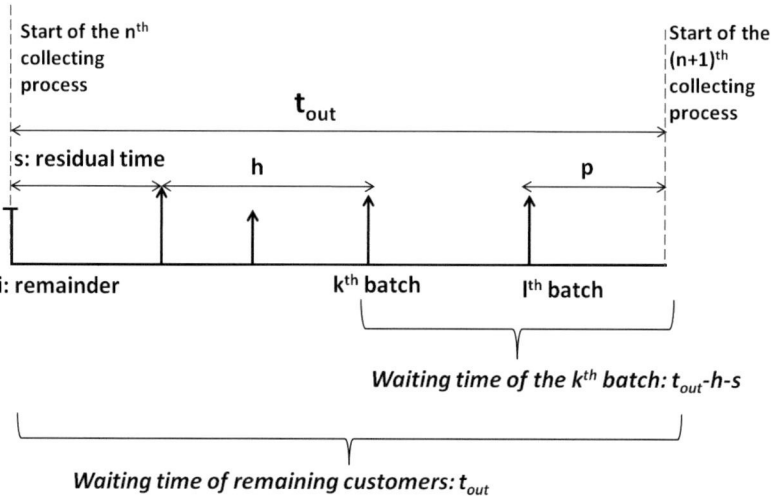

Figure 4.6.: Illustration of the waiting time of an arbitrary unit under the capacitated timeout rule, given that the collecting process finishes with a departing size $< K$

by equation 4.16. Note that $k = 0$ for the remaining transport units. for $k = 0$,

$$P(W^k = t_{out} \mid \widetilde{Y_{out}} < K) \approx \tag{4.15}$$

$$\sum_{s=1}^{a_{max}} \sum_{i=1}^{y_{max}-1} \sum_{l=1}^{l_{max}} \sum_{p=0}^{t_{out}-s} \sum_{j=1}^{K-i-1} r_{s,i} \cdot a_{t_{out}-s-p}^{(l-1)\otimes} \cdot \overline{a_{p+1}} \cdot y_{K-i-j}^{l\otimes} \cdot i$$

Obviously, this probability increases with the size of the remainder i. As a result, equation 4.16 is weighted by i.

If the transport unit does not belong to the remaining units, but arrives as a member of the kth batch where $1 \leq k \leq l_{max}$, the variable h is used to describe the time between the first arrival and the arrival of the kth batch. In this case the transport unit waits $(t_{out} - h - s)$ time units (see figure 4.6). The proportional value for this probability is given by for $1 \leq k \leq l_{max}$,

$$P(W^k = t_{out} - h - s \mid \widetilde{Y_{out}} < K) \approx \tag{4.16}$$

$$\sum_{s=1}^{a_{max}} \sum_{i=0}^{y_{max}-1} \sum_{l=k}^{l_{max}} \sum_{p=0}^{a_{max}-1} \sum_{h=0}^{t_{out}-s-p} \sum_{j=1}^{K-i-1} r_{s,i} \cdot a_h^{(k-1)\otimes} \cdot$$

$$a_{t_{out}-s-h-p}^{(l-k)\otimes} \cdot \overline{a_{p+1}} \cdot y_{K-i-j}^{l\otimes} \cdot \frac{K-i-j}{l}$$

Similar to equation 4.16, the equation given above is also multiplied with the expected batch size of the kth batch. Assuming that $(K - i - j)$ units are collected in l arrivals, the expected batch size for all arrivals is equal to $(K - i - j)/l$.

Subsequently, the waiting time of an arbitrary customer is investigated for the case $Y_{out} = K$. At least K units must have been collected in the relevant collecting process and the collecting time, denoted by m, must be less than or equal to t_{out}. As mentioned beforehand, the process finishes with a batch arrival. This kind of collecting process is visualized by figure 4.7. We, again, assume l arrivals and an initial residual state of (s, i). Provided that the transport unit was left over from the previous collecting process, its waiting time corresponds to m time units. The proportional value for this probability is computed by

for $k = 0$,

$$P(W^k = m \mid \widetilde{Y_{out}} = K) \approx \sum_{s=1}^{a_{max}} \sum_{i=1}^{y_{max}-1} \sum_{m=s}^{t_{out}} \sum_{l=1}^{l_{m,max}} \sum_{j=1}^{K-i} r_{s,i} \cdot$$

$$a_{m-s}^{(l-1)\otimes} \cdot y_{K-i-j}^{(l-1)\otimes} \cdot \overline{y_j} \cdot i \tag{4.17}$$

In the case that the transport unit arrives within the kth batch with $1 \leq k \leq l$, we again use the variable h as the time between the first arrival and the arrival of the kth batch. Consequently, the transport unit waits $(m - h - s)$ time units. In this analysis, we assume that $(K - i - j)$ units are collected in $(l - 1)$ arrivals, the expected batch size for all arrivals excluding the last batch is equal to $(K - i - j)/(l - 1)$. In accordance with this, the proportional probability must be weighted with $(K - i - j)/(l - 1)$ for $k = 1, 2, \cdots, (l - 1)$. On the other hand, the

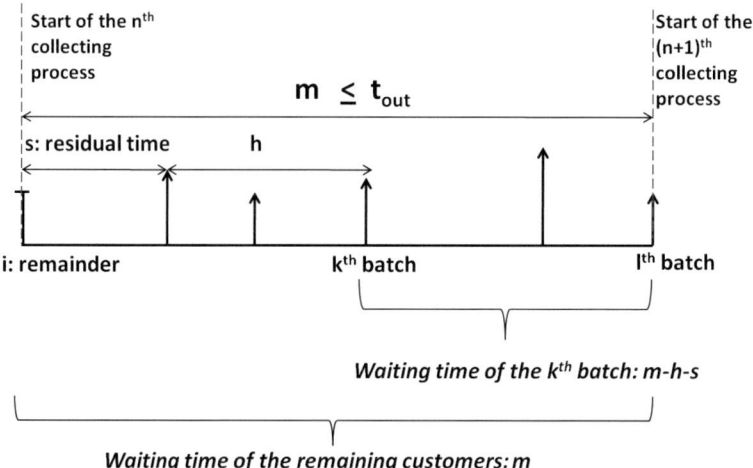

Figure 4.7.: Illustration of the waiting time of an arbitrary unit under the capacitated timeout rule, given that the collecting process finishes with a departing size $= K$

last batch has a size of $(n + j)$ units, of which just j depart and n are left over for the next process. Thus, the equation must be multiplied with j, which is the number of units that belong to the last batch and depart immediately without waiting. That's why, the variable V_{kl} is introduced to account for the special case of the last batch. It yields: for $1 \leq k \leq l_{max}$,

$$P(W^k = m - s - h \mid \widetilde{Y_{out}} = K) \approx \qquad (4.18)$$

$$\sum_{s=1}^{a_{max}} \sum_{i=0}^{y_{max}-1} \sum_{m=s}^{t_{out}} \sum_{l=k}^{l_{m,max}} \sum_{h=0}^{m-s} r_{s,i} \cdot a_h^{(k-1)\otimes} \cdot a_{m-h-s}^{(l-k)\otimes} \cdot$$

$$\sum_{j=1}^{K-i} \sum_{n=0}^{y_{max}-j} y_{K-i-j}^{(l-1)\otimes} \cdot y_{n+j} \cdot \left(V_{kl} \cdot j + (1 - V_{kl}) \cdot \frac{K - i - j}{l - 1} \right)$$

where

$$
V_{kl} = \begin{cases} 1 & \text{if } k = l \\ 0 & \text{otherwise} \end{cases} \tag{4.19}
$$

Subsequently, we normalize the proportional values, which are now denoted by $P^*(W^k = i \mid Y_{out} < K)$ and $P^*(W^k = i \mid Y_{out} = K)$.

$$
P(W = i \mid \widetilde{Y_{out}} < K) = \frac{\sum_{k=0}^{l_{max}} P^*(W^k = i \mid \widetilde{Y_{out}} < K)}{\sum_{j=0}^{t_{out}} \sum_{k=0}^{l_{max}} P^*(W^k = j \mid \widetilde{Y_{out}} < K)} \tag{4.20}
$$

$$
P(W = i \mid \widetilde{Y_{out}} = K) = \frac{\sum_{k=0}^{l_{max}} P^*(W^k = i \mid \widetilde{Y_{out}} = K)}{\sum_{j=0}^{t_{out}} \sum_{k=0}^{l_{max}} P^*(W^k = j \mid \widetilde{Y_{out}} = K)} \tag{4.21}
$$

Finally, the waiting time distribution is determined as follows:

$$
\begin{aligned}
P(W = i) \quad = \quad & P(W = i \mid \widetilde{Y_{out}} < K) \cdot P(\widetilde{Y_{out}} < K) \\
& + P(W = i \mid \widetilde{Y_{out}} = K) \cdot P(\widetilde{Y_{out}} = K)
\end{aligned} \tag{4.22}
$$

4.2.5. Numerical Results

Analysis of the effect of discretization

The introduced solution for the analysis of the capacitated timeout rule is exact given that the inter-arrival distribution is perfectly discrete. Nevertheless, the distribution of the inter-arrival time may be a continuous distribution in reality. When this is the case, the continuous distribution has to be discretized before the presented methods can be implemented. In this section, we study the effect of discretization on the quality of the results. For this purpose, we analyze a collecting process, in which the inter-arrival time is actually exponentially distributed with $\lambda = 0.2$. Other system figures are summarized in table 4.2.

We assume here that a number of observations of the inter-arrival time is available to the planner, that is sufficient to represent the population.

i	y_i	
1	0.9	$t_{out} : 100$
2	0.1	$K : 16$

Table 4.2.: System configuration for the analysis of the effect of discretization

To mimic this situation, we generated a vast number of exponentially distributed random numbers. It is assumed that the planner does not have the statistical background to realize/prove that the actual distribution of the inter-arrival time corresponds to an exponential distribution and computes an approximate discrete distribution based on the available data. Assuming an incremental time interval of length Δt, we obtain the discretized distribution as explained in Arnold and Furmans (2009, pp. 18). Thus,

$$a_i = P(A = i \cdot \Delta t) = P((i - 1) \cdot \Delta t < A \leq i \cdot \Delta t) \qquad (4.23)$$

In order to analyze the effect of discretization, we derived the waiting time distributions analytically under the assumptions of different Δt-values. To achieve this, the waiting time distributions for Δt-values 1, 2, and 5 are determined in discrete time domain. After that, we compared the resulting distributions with the actual waiting time distribution attained by simulation[1]. Figure 4.8 visualizes the probability mass function for the discrete approach with $\Delta t = 1$ in comparison to the actual density function. From the figure, it is clear that the discrete approach with $\Delta t = 1$ produces high quality results and matches the actual distribution very well.

Additionally, we investigated the discrete approaches with $\Delta t = 2$ and $\Delta t = 5$. In table 4.3, we display the mean and the quantiles of the waiting time under these cases. The table shows clearly that for high values of Δt, the results deviate significantly from simulation results.

The reason is that discretization aggregates the probabilities of all values in unit time interval. In this way, probabilities are computed. When

---

[1]The simulation tool eM-Plant was used.

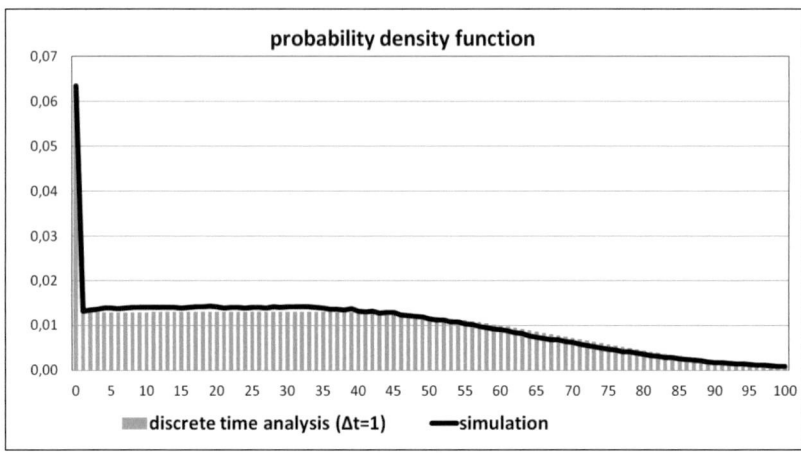

Figure 4.8.: Analysis of the effect of discretization: evaluation of the waiting time under the capacitated timeout rule for $\Delta t = 1$ in comparison to simulation results

the unit time interval is large, the approximate discrete distribution fails to reflect important characteristics of the actual inter-arrival time distribution. That's why, the resulting waiting time distribution becomes just a rough approximation of the actual distribution. Concluding, it should be stated that discrete time analysis gives near perfect results for smaller values of Δt. But the decision on Δt should be done carefully, as the results may suffer under higher values of Δt.

4.3. Batch Building: Capacity Interval Rule

In this section, we introduce the batch building rule "Capacity Interval Rule". In accordance with this rule, two parameters are employed to control the batch building process -namely, minimum and maximum collecting sizes. In reality, a collecting station has a limited capacity, which corresponds here to the maximum collecting size. Considering only the operational costs, the most efficient strategy is to collect the entities until the given capacity is fully exploited. On the other hand,

Approach	mean	95%-quantile	97%-quantile	99%-quantile
simulation	34.154	77	83	92
discrete analysis ($\Delta t = 1$)	36.081	79	85	93
discrete analysis ($\Delta t = 2$)	39.802	86	90	96
discrete analysis ($\Delta t = 5$)	46.134	90	95	96
$\Delta_{rel.}$ for ($\Delta t = 1$)	0.056	0.026	0.024	0.011
$\Delta_{rel.}$ for ($\Delta t = 2$)	0.165	0.117	0.084	0.043
$\Delta_{rel.}$ for ($\Delta t = 5$)	0.351	0.169	0.145	0.043

Table 4.3.: Analysis of the effect of discretization under the capacitated timeout rule: mean value and quantiles of the waiting time for $\Delta t = 1$, $\Delta t = 2$, and $\Delta t = 5$ compared to simulation results

such a strategy results in increased waiting time for the collected entities, thus, increases the inventory costs. Often, a minimum collecting size is defined, above which the collected batch is allowed to depart. In this way, waiting time is reduced.

This kind of batch building is applied frequently in transport systems, in which a minimum degree of utilization for the employed vehicles is required. The capacity of the vehicle corresponds then to the maximum collecting size. Under this rule, transport process takes place as soon as the minimum number of transport units is collected.

4.3.1. Queuing System

In this analysis, the minimum and the maximum collecting sizes are denoted by L and K, respectively. We assume that the transport units arrive at the collecting station in stochastic batch sizes (Y) and in stochastic time intervals (A). The variables A and Y are assumed to be iid. Figure 4.9 illustrates the collecting process under the capacity interval rule. For this batch building mode, we present exact solutions for the departing batch size Y_{out}, inter-departure time D_{out}, and the waiting time of an arbitrary unit W. The list of important variables and parameters used in this analysis is given below.

L minimum collecting size,
K maximum collecting size,
A inter-arrival time,
Y batch size of an incoming batch,
R^y remainder, number of transport units left over from the previous collecting process due to the limited collecting size,
N number of arrivals in a collecting process,
Y_{col} number of transport units collected in a collecting process,
Y_{out} departing batch size,
D_{out} inter-departure time between two successive departures,
W waiting time of an arbitrary transport unit.

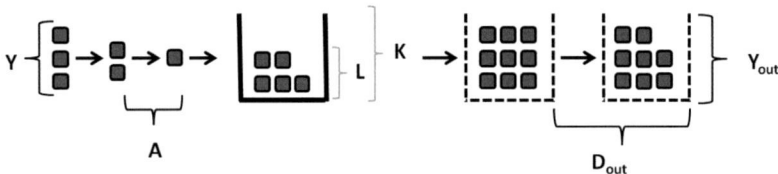

Figure 4.9.: Collecting process under the capacity interval rule

We assume in our analysis, that at least one arrival is needed to complete the collecting process. In other words, the number of remaining units from the previous collecting process is never enough to satisfy the minimum collecting size. Therefore, it is assumed that $y_{max} \leq L$. Under this assumption, multiple departures cannot occur at the same time instant. Finally, we again assume that $y_0^{0\otimes} = a_0^{0\otimes} = 1$ and process the events in the following sequence: 1) arrival 2) end of a collecting process 3) start of a collecting process.

Remainder distribution

In this model, every collecting process finishes with a batch arrival, hence, the residual inter-arrival time corresponds to a complete inter-arrival time and does not have to be computed additionally. On the other hand, some units have to wait for the subsequent collecting process due to the limited collecting size. The remainder (R^y) stands for the number of remaining customers and its distribution is needed to derive the performance measures.

The remainder of the nth process depends on the remainder of the $(n-1)^{th}$ process but not on the $(n-2)^{th}$ process. Consequently, it is possible to derive the remainder distribution with the aid of a discrete Markov chain. Like the case in section 4.2.1, the Markov chain is ergodic, thus, the steady state distribution of the remainder exists for $n \to \infty$.

In order to compute the transition probability $p(i, m)$, i.e. transition probability from an initial remainder of i to m units, we define the interim variable Y_{col}, which represents the number of collected units immediately before a departure instant. The maximum value of this variable is limited to $(L + y_{max} - 1)$ since the number of collected units immediately before the last batch must be less than L and the last batch may have a maximum size of y_{max}. The conditional distribution of Y_{col} is computed as follows.

for $n = 0, \cdots, y_{max} - 1$,

$$P(Y_{col} = L + n \mid R^y = i) = \sum_{l=1}^{l_{max}} \sum_{j=1}^{L-i} y_{L-i-j}^{(l-1)\otimes} \cdot y_{j+n} \qquad (4.24)$$

where l_{max} is the maximum number of arrivals that may take place in a collecting process and computed with the following expression for the rest of our analysis.

$$l_{max} = \lceil \frac{L}{y_{min}} \rceil \qquad (4.25)$$

The remainder becomes zero if the number of collected units is less than or equal to K.

for $m = 0$,

$$p(i, m) = \sum_{m=0}^{K-L} P(Y_{col} = L + m \mid R^y = i) \qquad (4.26)$$

The remainder assumes a positive value if the number of collected transport units is greater than K:

for $m \neq 0$,

$$p(i, m) = P(Y_{col} = K + m \mid R^y = i) \qquad (4.27)$$

51

Having computed the transition probabilities, we define the stationary equations as follows:

$$r_m^y = \sum_{i=0}^{r_{max}^y} p(i,m) \cdot r_i^y \qquad (4.28)$$

where r_{max}^y is the maximum value of the remainder and given with the following expression:

$$r_{max}^y = \max\{L + y_{max} - 1 - K, 0\}$$

Finally, an additional equation arises due to the fact that the sum of steady state probabilities is equal to one.

$$\sum_{i=0}^{i_{max}} r_m^y = 1 \qquad (4.29)$$

As the case with the capacitated timeout rule, we attain an overdetermined equation system, thus, one equation must be omitted.

In the case that $L + y_{max} - 1 \leq K$, the maximum number of collected units can not exceed the maximum collecting size K. Consequently, the remainder is always equal to zero in this particular case:

for $L + y_{max} - 1 \leq K$,

$$r_0^y = 1 \qquad (4.30)$$

4.3.2. Departing Batch Size Distribution

Having computed the conditional probability $P(Y_{col} = L + n \mid R^y = i)$, we are now capable of deriving the distribution of Y_{col} based on the law of total probability

for $n = 0, 1, \cdots, y_{max} - 1$,

$$P(Y_{col} = L + n) = \sum_{i=0}^{r_{max}^y} P(Y_{col} = L + n \mid R^y = i) \cdot r_i^y \qquad (4.31)$$

The departing batch size distribution y_{out} corresponds to the distribution y_{col} with an upper bound K. Thus, we use the operator Π^M in equation 2.13 and obtain the departing batch size distribution:

$$y_{out} = \Pi^K[y_{col}] \tag{4.32}$$

4.3.3. Inter-departure Time Distribution

As the first step, we derive the number of arrivals needed to complete a collecting process.

for $l = 1, \cdots, l_{max}$,

$$P(N = l) = \sum_{i=0}^{r_{max}^y} \sum_{m=1}^{L-i} \sum_{n=0}^{y_{max}-m} r_i^y \cdot y_{L-i-m}^{(l-1)\otimes} \cdot y_{m+n} \tag{4.33}$$

where l_{max} is computed in accordance with equation 4.25. And finally, we obtain the inter-departure time distribution as follows:

for $i = 1, \cdots, l_{max} \cdot a_{max}$,

$$P(D_{out} = i) = \sum_{l=1}^{l_{max}} P(N = l) \cdot a_i^{l\otimes} \tag{4.34}$$

4.3.4. Waiting Time Distribution

For the derivation of the waiting time, we represent here an analogous procedure like the one for the capacitated timeout rule (see section 4.2.4). As the initial step, we derive the probability, that an arbitrary transport unit departs after a collecting process, within which l arrivals took place. This probability increases with the increasing departing batch size. Regarding this fact, we multiply the succeeding equation with the relevant departing batch size and normalize the expression by the expected value of the departing batch size:

53

$$P(\tilde{N} = l) = \sum_{i=0}^{r_{max}^y} \sum_{m=1}^{L-i} r_i^y \cdot y_{L-i-m}^{(l-1)\otimes} \cdot \qquad (4.35)$$

$$\left(\sum_{n=0}^{K-L} y_{m+n} \cdot \frac{L+n}{E(Y_{out})} + \sum_{K-L+1}^{y_{max}-1} y_{m+n} \cdot \frac{K}{E(Y_{out})} \right)$$

Following this, we analyze the waiting time of an arbitrary transport unit given the number of arrivals observed within the collecting process (N). For the case $N = l$, we assume that the transport unit belongs to an arbitrary k^{th} batch with $k : 0, 1, \cdots, l$. As the batch 0, we denote the units that were left over from the preceding collecting process. These units have to wait a complete collecting time.

for $k = 0$, we obtain:

$$P(W^{k,l} = m \mid \tilde{N} = l) \approx \sum_{i=0}^{r_{max}^y} \sum_{n=1}^{L-i} \sum_{p=0}^{y_{max}-n} \sum_{m=l}^{l \cdot a_{max}} r_i^y \cdot i \cdot y_{L-n-i}^{(l-1)\otimes} \cdot y_{n+p} \cdot a_m^{l\otimes}$$

$$(4.36)$$

Note that the expression is weighted by the remainder (i).

Subsequently, we presume that the transport unit arrives as the member of the kth batch where $k : 1, \cdots, (l-1)$. Assuming that $(L - n - i)$ transport units are collected in $(l-1)$ arrivals, we again multiply the proportional expression with the expected size of the batch, which is given by $\frac{L-n-i}{l-1}$ in the following equation. Thus, we attain

for $k = 1, \cdots, (l-1)$,

$$P(W^{k,l} = m \mid \tilde{N} = l) \approx \sum_{i=0}^{r_{max}^y} \sum_{n=1}^{L-i} \sum_{p=0}^{y_{max}-n} \sum_{m=l-k}^{(l-k) \cdot a_{max}} r_i^y \cdot$$

$$y_{L-n-i}^{(l-1)\otimes} \cdot \frac{L-n-i}{l-1} \cdot y_{n+p} \cdot a_m^{(l-k)\otimes} \quad (4.37)$$

Eventually, we examine the waiting time for the case, that the unit belongs to the last batch ($k = l$). Note that some units of the last batch may not depart immediately owing to the capacity restriction. That's why, the expression for this case is not weighted with the batch size of the last batch but rather with the number of the units of the last batch that depart immediately. It yields:

for $k = l$,

$$P(W^{k,l} = 0 \mid \widetilde{N} = l) \approx \sum_{i=0}^{r^y_{max}} \sum_{n=1}^{L-i} r^y_i \cdot y^{(l-1)\otimes}_{L-n-i} \cdot \tag{4.38}$$

$$\left(\sum_{p=0}^{K-L} y_{n+p} \cdot (n+p) + \sum_{p=K-L+1}^{y_{max}-1} y_{n+p} \cdot (K-L+n) \right)$$

We derived by means of these expressions the proportional values for $P(W^{k,l} = s \mid \widetilde{N} = l)$. Let us denote these proportional expressions with $P^*(W^{k,l} = s \mid \widetilde{N} = l)$. Normalizing these values, we obtain

$$P(W = s \mid \widetilde{N} = l) = \frac{\sum_{l=1}^{l_{max}} \sum_{k=0}^{l} P^*(W^{k,l} = s \mid \widetilde{N} = l)}{\sum_{s=0} \sum_{l=1}^{l_{max}} \sum_{k=0}^{l} P^*(W^{k,l} = s \mid \widetilde{N} = l)} \tag{4.39}$$

Finally, we obtain the waiting time distribution by

$$P(W = s) = \sum_{N=1}^{l_{max}} P(W = s \mid \widetilde{N} = l) \cdot P(\widetilde{N} = l) \tag{4.40}$$

4.4. Batch Queue: $G^X/G^{[L,K]}/1$-Queue

In the previous sections, we focused on the batch building processes, in which collecting process finishes upon the fulfillment of a predefined control strategy. Such control strategies are often parameterized by the minimum or the maximum values of the collecting size or the collecting time. Particularly, we analyzed the capacity interval rule. Complying with this mode of batch building, a minimum collecting size is needed to complete a collecting process.

In the current section, we examine a batch server queue, in which a batch service process with a stochastic duration is performed. Kendall's notation for the considered batch queue is $G^X/G^{[L,K]}/1$. Similar to the capacity interval rule, a minimum number of units (L) has to be collected, in order to trigger a service process. Besides, the number of units, that can be served simultaneously, is bounded to the server capacity (K).

As mentioned in section 4.3, the capacity interval rule is often implemented in transport systems to assure a minimum utilization for the vehicles. Furthermore, we explained that a new vehicle becomes available immediately after the departure of the previous one in such a transport system. In contrast, a transport system can be modeled by means of a $G^X/G^{[L,K]}/1$-queue, e.g. if there is one vehicle, that shuttles between two destinations and may not be available each time the minimum utilization is reached. As a result of this, a departure may take place, not only when the minimum collecting size is fulfilled, but also when a vehicle is available.

4.4.1. Queuing System

As in the previous sections, we assume that transport units arrive at the system in stochastic batch sizes and in stochastic time intervals, denoted by Y and A, respectively. Moreover, the service process is characterized by the service time B. We assume that these random variables are iid.

In the $G^X/G^{[L,K]}/1$-queue, incoming batches are collected throughout the service time. Whenever a service process finishes, the number of waiting units is observed. In the case that the number of waiting units is less than L, then the server becomes idle until at least L units are collected in the queue. For the case that there are $\geq L$ and $\leq K$ units in the queue, all the units are served as a batch. And finally if there are $> K$ units in the queue, only K of them are served immediately. The remaining units will be served in the subsequent service processes.

We present here methods to compute the distributions of the departing batch size (Y_{out}), inter-departure time (D_{out}), and the waiting time (W). The list of important parameters and variables used in the analysis is given below.

L minimum number of units to trigger a service process,
K capacity of the batch server,

B	service time,
A	inter-arrival time,
Y	batch size of an incoming batch,
R^y	remainder, number of units left in the queue after the start of a service process due to the capacity restriction,
R^a	residual inter-arrival time, time interval between the start of a service process and the first arrival after the start of the service,
R	residual state, two dimensional variable defining the states of R^a and R^y at the beginning of a service start,
r^y_{max}	maximum value of the remainder,
Y_{col}	queue length immediately before the start of a service,
Y_{out}	departing batch size,
IT	idle time,
D_{out}	inter-departure time between two departures,
W	waiting time of an arbitrary unit.

The queuing system is illustrated in figure 4.10.

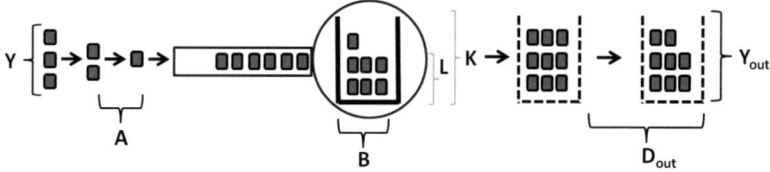

Figure 4.10.: Illustration of the $G^X/G^{[L,K]}/1$-batch server queue

Similar to our previous assumptions, we assume that at least one arrival occurs within a service time. To assure this, the assumptions $a_{max} \leq b_{min}$ and $y_{max} \leq L$ are made. Likewise, we assume that zero-fold convolutions of the distributions a, b, and y are Dirac distributed with a constant value of zero. Thus, $y_0^{0\otimes} = a_0^{0\otimes} = b_0^{0\otimes} = 1$. In this analysis, we assume that the system is in a steady state. The queue system is stable, if the mean number of collected units within a service time is less than the server capacity; $\frac{E[B] \cdot E[Y]}{E[A]} < K$. Concluding, we assume that events are processed in the following sequence: 1) arrival 2) end of service 3) start of service.

Residual state distribution

Owing to the limited capacity of the batch server, some units may be left in the queue at the start of a service process. As in the previous section, the remainder refers here again to the number of remaining units at the start of a service process. Moreover, the time interval between the start of a service process and the first arrival within the service time can be a complete inter-arrival time or a residual of it. Complying with our preceding analysis, this time interval is represented by the residual inter-arrival time. Similar to the argumentation in section 4.2.1, we employ the residual state (R) as the two dimensional variable defining the states of the residual inter-arrival time and the remainder simultaneously. Since the residual state at the beginning of nth service process depends only on that of the $(n-1)^{th}$ service process, we identify a Markov process. Analogous to the collecting process under the capacitated timeout rule, the Markov chain is ergodic and the joint distribution of the residual time and the remainder can be determined as the steady state distribution.

Let us investigate the possible values of the residual time and the remainder. We consider now an arbitrary n^{th} service process. At the end of the service time, the queue length is inspected. If the queue length is less than L, the server becomes idle and waits for additional arrivals. This case is demonstrated in figure 4.11. Once at least L units are col-

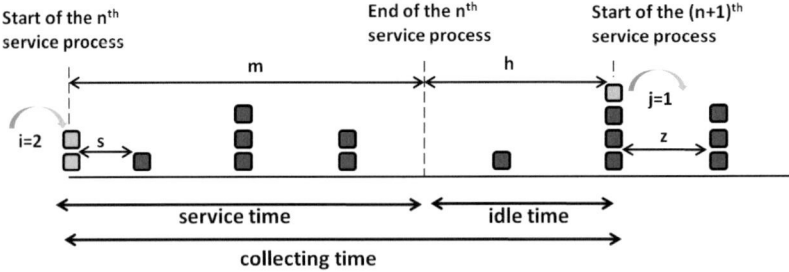

Figure 4.11.: Collecting process in the $G^X/G^{[L,K]}/1$-queue under case 1: idle time arises at the end of the service time

lected, then the idle time finishes and the $(n+1)^{th}$ service process starts. Thus, the nth collecting process comprises the nth service time and the nth idle time. Such a collecting process finishes always with a batch arrival. Hence, the residual time corresponds to a complete inter-arrival time. The number of collected units prior to the arrival of the last batch must be less than L. If the last batch completes the queue length to a value less than or equal to K, the whole queue is served in the $(n+1)^{th}$ service process. So the remainder becomes zero. Otherwise, only K units are served and the excessive units are left in the queue. If, for instance, $L = 10$ and $K = 12$ in the example depicted in figure 4.11, one unit will be left. As the maximum number of collected units is $(L + y_{max} - 1)$, the maximum value of the remainder corresponds to $(L + y_{max} - 1 - K) < y_{max}$ in this case. Consequently, the maximum value of the remainder is bounded to $(y_{max} - 1)$.

In the contrary case, which is depicted in figure 4.12, the queue length at the end of the nth service time is greater than or equal to L units. Therefore, the number of waiting units suffices to initiate the $(n + 1)^{th}$ service process immediately. In this case, the collecting time equates

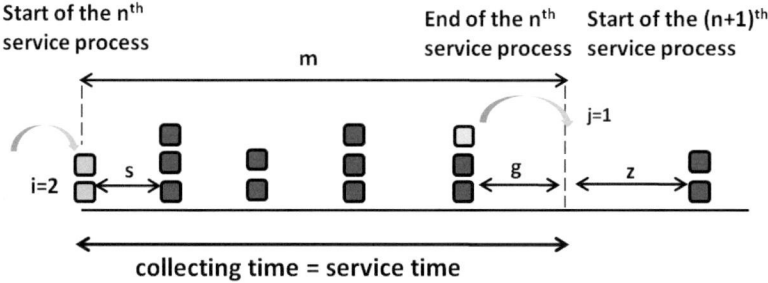

Figure 4.12.: Collection process in the $G^X/G^{[L,K]}/1$ under case 2: no idle time arises at the end of the service time

to the nth service time and does not have to end with the arrival of the last batch. Consequently, the residual time can also be a residual of the inter-arrival time and may assume any whole number from 1 to a_{max}. As in the first case, this kind of collecting process may yield some waiting units after the start of $(n + 1)^{th}$ service process. This happens, when the queue length at the end of the nth service time is greater than K. Different than all the cases, we investigated so far, the maximum value of the remainder does not have to be less than y_{max}. We denote the unknown maximum value of the remainder as r^y_{max}. Given that the system is stable, r^y_{max} must have a finite value.

Having investigated both cases, it follows that the state space of the residual time is defined by the set $S^a = 1, 2, \cdots, a_{max}$ whereas the state space of the remainder is the set $S^y = 0, 1, \cdots, r^y_{max}$. We build a Markov chain with $r^y_{max} \cdot a_{max}$ states and compute the residual state distribution with the following stationary equations.

$$r_{z,j} = \sum_{s=1}^{a_{max}} \sum_{i=0}^{r^y_{max}} p(s,i)(z,j) \cdot r_{s,i} \tag{4.41}$$

where $p(s,i)(z,j)$ stands for the transition probability from an initial residual state (s,i) to (z,j). A further equation results from the sum of steady state probabilities:

$$\sum_{z=1}^{a_{max}} \sum_{j=0}^{r^y_{max}} r_{z,j} = 1 \tag{4.42}$$

Subsequently, we get an overdetermined equation system with $a_{max} \cdot r^y_{max} + 1$ equations and one equation has to be discarded. Since the maximum value of the remainder is unknown, we have to truncate the method in accordance with the assumed value of r^y_{max}. Therefore, the method is exact within an ϵ-environment. Numerical results show that an overestimated r^y_{max} has no significant effect on the quality of the results, whereas the underestimation worsens the quality of the method (see section 4.4.5). For this reason, it is recommended to try iterative values of r^y_{max} and choose the minimum one, above which the joint distribution $r_{j,z}$ does not change significantly.

Referring to the cases introduced in this section, transition probabilities are calculated; i.e. transition probabilities are computed conditioned on the given case. Firstly, we derive the probability of attaining a residual state of (z,j) from an initial state of (s,i) conditioned on the case, in which idle time arises at the end of the service process.Following equation gives this probability

$$p^1(s,i)(z,j) = \sum_{m=b_{min}}^{b_{max}} \sum_{l=2}^{l_{max}} \sum_{h=1}^{l \cdot a_{max}-m} \sum_{p=1}^{L-i-1} b_m \cdot a_{m+h-s}^{(l-1)\otimes} \cdot a_z \cdot \quad (4.43)$$

$$y_{L-i-p}^{(l-1)\otimes} \cdot \left(VJ \sum_{d=0}^{K-L} y_{p+d} + (1-VJ) \cdot y_{K+j-L+p} \right)$$

where

$$VJ = \begin{cases} 1 & \text{if } j = 0 \\ 0 & \text{otherwise} \end{cases} \quad (4.44)$$

According to equation 4.44, the process starts with a remainder of i transport units and it takes s time units till the first arrival is observed. We assume here that the service time lasts m time units and an idle time of h time units is needed to complete the collecting process (see figure 4.11). The variable l denotes here the total number of arrivals in $(m+h)$ time units, of which maximum value is given by:

$$l_{max} = \lceil \frac{L}{y_{min}} \rceil \quad (4.45)$$

With the last batch, the collecting process finishes and the subsequent process starts. Hence, the time between the start of the next process and the first arrival corresponds to a complete inter-arrival time of z time units. The variable p defines the number of missing transport units after $(l-1)$ arrivals. So, the last batch must have a size of at least p transport units. If the resulting remainder (j) equals to zero, then the last batch must complete the queue length to value less than or equal to K. Otherwise, the last batch must have a batch size of $(K+j-L-p)$. To account for this, we introduced the variable VJ.

In the second case, at least L transport units are collected at the end of the service process and the next service process starts immediately. We compute the transition probability $p^2(s,i)(z,j)$ as follows:

$$p^2(s,i)(z,j) = \sum_{m=b_{min}}^{b_{max}} \sum_{l=1}^{l_{m,max}} \sum_{g=0}^{m-s} b_m \cdot a_{m-s-g}^{(l-1)\otimes} \cdot a_{g+z} \cdot$$

$$\left(VJ \sum_{x=L-i}^{K-i} y_x^{l\otimes} + (1-VJ) \cdot y_{K-i+j}^{l\otimes} \right) \tag{4.46}$$

In equation 4.46, we consider a service process, which starts with an initial residual state of (s,i) and takes m time units (see figure 4.12). We assume that l arrivals take place in m time units. The maximum value of l is conditioned on m:

$$l_{m,max} = \lceil \frac{m}{a_{min}} \rceil \tag{4.47}$$

As mentioned beforehand, this kind of collecting process does not have to finish with a batch arrival. The variable g denotes the time interval between the arrival of the last batch and the end of the service (collecting) process. The next arrival must take $(g+z)$ time units so that a residual time of z time units is observed. The variable VJ is also employed in this equation.

Eventually, transition probability $p(s,i)(z,j)$ is obtained by:

$$p(s,i)(z,j) = p^1(s,i)(z,j) + p^2(s,i)(z,j) \tag{4.48}$$

4.4.2. Departing Batch Size Distribution

In order to derive the departing batch size distribution (Y_{out}), we firstly compute the distribution of the queue length prior to the start of a service process (Y_{col}). For this purpose, we again differentiate between the cases, elucidated in the previous section. Under the condition that server becomes idle at the end of the service time, Y_{col} is determined with the following equation:

$$P^1(Y_{col} = L+d) = \sum_{z=1}^{a_{max}} \sum_{j=0}^{r_{max}^y} \sum_{m=b_{min}}^{b_{max}} \sum_{l=2}^{l_{max}} \sum_{h=1}^{l \cdot a_{max}-m} \sum_{p=1}^{L-j-1}$$

$$\sum_{d=0}^{y_{max}-p} r_{z,j} \cdot b_m \cdot a_{m+h-z}^{(l-1)\otimes} \cdot y_{L-j-p}^{(l-1)\otimes} \cdot y_{p+d} \quad (4.49)$$

In equation 4.49, a collecting process is investigated which starts with a residual state of (z,j). In other words, the residual inter-arrival time is z time units and there were already j transport units at the beginning of the service process. We assume that the service time takes m time units and the server stays for h time units idle at the end of the service time. The variable l refers here to the number of arrivals within $(m+h)$ time units and is upper bounded to l_{max} (see equation 4.45). In $(l-1)$ arrivals, $(L-j-p)$ units are collected. Thus, the queue length is $(L-p)$ immediately before the arrival of the last batch. Assuming a batch size of $(p+d)$ units for the last batch, Y_{col} becomes $(L+d)$ units.

In the following equation, we address the case, in which collecting process finishes with the service time. In a similar fashion like the first case, we again assume an initial residual state of (z,j) and a service time of m units. Suppose that l arrivals take place in the service time and insert x transport units to the queue. Accordingly, Y_{col} is equal to $(j+x)$ at the end of the service time.

$$P^2(Y_{col} = j+x) = \sum_{z=1}^{a_{max}} \sum_{j=0}^{r_{max}^y} \sum_{m=b_{min}}^{b_{max}} \sum_{l=1}^{l_{m,max}} \sum_{g=0}^{m-z} \sum_{x=L-j}^{y_{max}\cdot l} r_{z,j} \cdot \quad (4.50)$$

$$b_m \cdot a_{m-z-g}^{(l-1)\otimes} \cdot \overline{a_{g+1}} \cdot y_x^{l\otimes}$$

Recall that the maximum value of l in this case is $l_{m,max}$ and can be determined by means of equation 4.47. Finally, it yields:

$$P(Y_{col} = i) = P^1(Y_{col} = i) + P^2(Y_{col} = i) \quad (4.51)$$

The distribution of the departing batch size corresponds to the distribution of Y_{col} with an upper bound K. So as to determine the distribution of a RV with an upper bound, we use the operator introduced in equation 2.13. Consequently, the distribution y_{out} is obtained:

$$y_{out} = \Pi^K [y_{col}] \tag{4.52}$$

4.4.3. Inter-departure Time Distribution

So as to compute the inter-departure time distribution, we derive firstly the idle time distribution. The idle time at the end of an nth service process (IT^n) depends on the service time observed in this process (B^n). That's why, we derive the idle time distribution, given the service time. Following conditional probabilities are computed.

for h: $1, 2, \cdots, h_{max} = \lceil \frac{L-1}{y_{min}} \rceil \cdot a_{max}$,

$$P(IT = h \mid B = m) = \sum_{z=1}^{a_{max}} \sum_{j=0}^{r^y_{max}} \sum_{l=2}^{l_{max}} \sum_{p=1}^{L-j-1} r_{z,j} \cdot y_{L-j-p}^{(l-1)\otimes} \cdot \overline{y_p} \cdot a_{m+h-z}^{(l-1)\otimes} \tag{4.53}$$

The probability, that no idle time arises at the end of a service process of length m units, results from the sum of steady state probabilities.

$$P(IT = 0 \mid B = m) = 1 - \sum_{h=1}^{h_{max}} P(IT = h \mid B = m) \tag{4.54}$$

Based on the law of total probability, the idle time distribution can be computed now:

$$P(IT = i) = \sum_{m=b_{min}}^{b_{max}} P(IT = i \mid B = m) \cdot b_m \tag{4.55}$$

As illustrated in figure 4.13, the nth inter-departure time interval consists of the idle time after the previous service process (IT^{n-1}) and the service time of the given service process (B^n). As these components

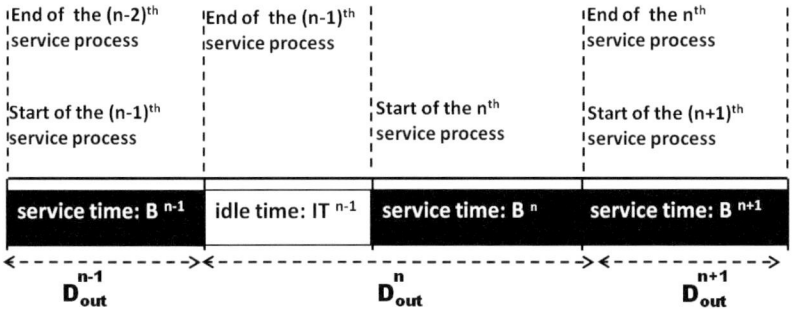

Figure 4.13.: Analysis of the inter-departure time for the $G^X/G^{[L,K]}/1$-queue

are independent of each other, we can obtain the inter-departure time distribution as the convolution of the idle time and the service time distributions. Hence,

$$dout = it \otimes b \tag{4.56}$$

4.4.4. Waiting Time Distribution

For the analysis of the waiting time of an arbitrary transport unit, we develop a similar approach as introduced in section 4.2.4. For this purpose, we fall back on the cases introduced in section 4.4.1. Firstly, we derive the probability that the observed transport unit arrives throughout a collecting process, in which idle time arises at the end of the service process. The probability for this case is proportional to:

$$\widetilde{P(C1)} = \sum_{z=1}^{a_{max}} \sum_{j=0}^{r_{max}^y} \sum_{m=b_{min}}^{b_{max}} \sum_{l=2}^{l_{max}} \sum_{h=1}^{l \cdot a_{max}-m} \sum_{p=1}^{L-j-1} \sum_{d=0}^{y_{max}-p} r_{z,j} \cdot b_m \cdot$$
$$a_{m+h-z}^{(l-1)\otimes} \cdot y_{L-j-p}^{(l-1)\otimes} \cdot y_{p+d} \cdot (L-j+d) \tag{4.57}$$

Obviously, this probability increases with the increasing number of units, that arrive within such a process. As a result, the relation above is multiplied with $(L - j + d)$.

As the next step, we determine the probability that an arbitrary transport unit arrives within a service process, at the end of which no idle time occurs. Similar to the equation (4.57), this probability is also multiplied with the number of units that arrived within the service time.

$$\widetilde{P(C2)} = \sum_{z=1}^{a_{max}} \sum_{j=0}^{r^y_{max}} \sum_{m=b_{min}}^{b_{max}} \sum_{l=1}^{l_{m,max}} \sum_{g=0}^{m-z} \sum_{x=L-j}^{y_{max} \cdot l} r_{z,j} \cdot \tag{4.58}$$
$$b_m \cdot a_{m-z-g}^{(l-1)\otimes} \cdot \overline{a_{g+1}} \cdot y_x^{l\otimes} \cdot x$$

We introduce the variable Y_{arr} as the number of units that arriving within a collecting process (thus, $Y_{arr} = Y_{col} - R^y$). Summing up these proportional values, we calculate the expected value of Y_{arr},

$$E(Y_{arr}) = \widetilde{P(C1)} + \widetilde{P(C2)}$$

As we weight the expressions for $\widetilde{P(C1)}$ and $\widetilde{P(C2)}$ with the number of units arrived within the collecting process, we need to normalize them by dividing the expressions with $E(Y_{arr})$. In this way, we are able to derive the exact expressions for $P(C1)$ and $P(C2)$.

$$P(C1) = \frac{\widetilde{P(C1)}}{E(Y_{arr})} \qquad P(C2) = \frac{\widetilde{P(C2)}}{E(Y_{arr})}$$

Now, we study the waiting time of an arbitrary unit within an arbitrary collecting process conditioned on these cases.

Case 1 At first, we investigate the waiting time of an arbitrary transport unit that belongs to an arbitrary k^{th} batch, under the first case. This kind of process finishes with a batch arrival and the number of collected units prior to the last batch is always less than L. All the collected units that do not belong to the last batch are served in the next service process. For the units of the last batch (l^{th} batch), a distinction has to be made, as some members of the last batch may not be served in the next service process. This occurs, when the last batch completes the queue length to a value, that is greater than K. Moreover, all the batches prior to the last batch have the same expected size. In order to distinguish the last batch, we define the variable KL:

$$KL = \begin{cases} 1 & \text{if} \quad k = l \\ 0 & \text{otherwise} \end{cases} \tag{4.59}$$

The waiting time under this case is then proportional to:

$$P(W = m + h - u + KL \cdot d \mid C1) \approx \tag{4.60}$$

$$\sum_{z=1}^{a_{max}} \sum_{j=0}^{r_{max}^y} \sum_{m=b_{min}}^{b_{max}} \sum_{l=2}^{l_{max}} \sum_{h=1}^{l \cdot a_{max} - m} \sum_{k=1}^{l} \sum_{u=z}^{m+h} \sum_{p=1}^{L-j-1} r_{z,j} \cdot$$

$$b_m \cdot a_{u-z}^{(k-1)\otimes} \cdot a_{m+h-u}^{(l-k)\otimes} \cdot y_{L-j-p}^{(l-1)\otimes} \cdot$$

$$\left(KL \sum_{n=p}^{y_{max}} \sum_{x=1}^{n} \sum_{d=0}^{b_{max} \cdot F} y_n \cdot b_d^{F\otimes} + (1 - KL) \cdot \overline{y_p} \cdot \frac{L - j - p}{l - 1} \right)$$

where $F = \lceil \frac{L-p+x}{K} \rceil - 1$.

We investigate with the above equation possible courses of a collecting process conditioned on the first case. The variable m represents the service time and the variable h is the idle time. We assume that there are l arrivals within $(m+h)$ time units and l_{max} is the maximum number of arrivals within the collecting process (see equation 4.45). The variable u is the time between the start of the service process and the arrival of the kth batch. These variables are illustrated in figure 4.14.

Let us consider the case that k is different than l. In this case, the number of collected units till the arrival of the kth batch is always less than L. Accordingly, units of the kth batch are served in the succeeding service period. Therefore, the units that belong to the kth batch wait only $(m+h-u)$ time units. Since all the batches prior to the last batch have the same expected batch size, the equation is weighted by $(\frac{L-j-p}{l-1})$.

Under case 1, the number of collected units after the arrival of the last batch may exceed K. Consequentially, a transport unit that belongs to

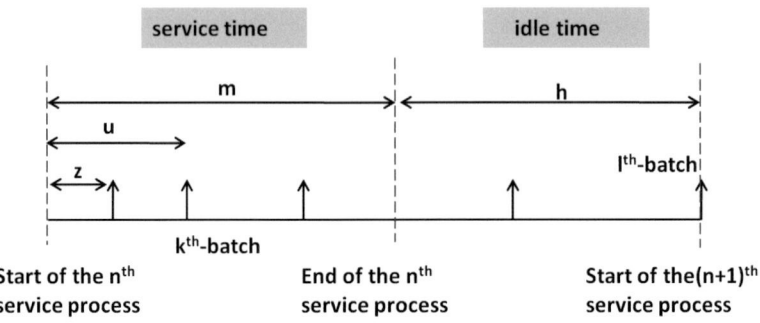

Figure 4.14.: Analysis of the waiting time of an arbitrary unit in the $G^X/G^{[L,K]}/1$-queue under case 1

the last batch ($k = l$) may wait a complete service time. In this case $u = m + h$ and the variable d stands for the additional complete service time. In order to detect if the unit waits an additional service period, we adopt a slightly different approach than the approach for other batches prior to the last batch. In accordance with this approach, we determine if the unit has to wait a complete service process or served immediately. The variable F stands for the number of complete service times that the unit has to wait. In order to obtain F, we determine firstly the position of the unit as the sum of all collected units prior to the arrival of last batch and the position of the relevant transport unit in the last batch, thus, it is given by $(L-p+x)$ in the previous equation. If the position of the transport unit is $> K$ and $\leq 2 \cdot K$, F becomes one and the unit has to wait a complete service period. In the contrary case, if the position of the transport unit $\leq K$, F becomes zero. Thus, d must be zero. Note that the transport units can wait maximum one complete service period, as the maximum number of collected units $L + y_{max} - 1$ under this case is always less than $2 \cdot K$. In other words, F can be maximum one under case 1.

Case 2 In the second case, no idle time occurs. We assume that the service time equals m and the number of arrivals within the service time is equal to l. We denote the maximum number of arrivals conditioned on a service time of m time units by $l_{m,max}$ (see equation 4.47). The waiting time under this case has two components. Firstly, the first component is the waiting time until the start of the next service process. Assuming that the time between the start of the previous service process and the arrival of the kth equals u, this waiting time amounts to $(m - u)$ and is the same for all the units that belong to the kth batch (see figure 4.15).

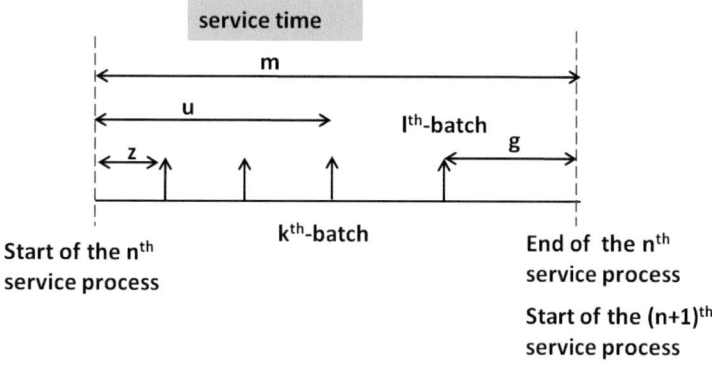

Figure 4.15.: Analysis of the waiting time of an arbitrary unit in the $G^X/G^{[L,K]}/1$-queue under case 2

The second component of the waiting time is a multiple of complete service times that the unit may wait additionally and depends on the position of the unit in the queue. The variable d stands for this kind of waiting and is simply the duration of all complete service processes that the individual transport unit has to wait additionally. To compute d, the position of the unit is assessed as the sum of all collected units prior the kth batch and the position of the unit in the kth batch, given with the expression $(j + n + t)$. If the observed unit is waiting e.g. in the sixth position in the queue and $K = 4$, then the unit has to wait one additional service time. The variable F is used to calculate the number of complete service processes that the unit has to wait additionally. The waiting time under the second case is, therefore, proportional to:

$$P(W = m - u + d \mid C2) \approx \qquad (4.61)$$

$$\sum_{z=1}^{a_{max}} \sum_{j=0}^{r^y_{max}} \sum_{m=b_{min}}^{b_{max}} \sum_{l=1}^{l_{m,max}} \sum_{k=1}^{l} \sum_{u=z}^{m} \sum_{g=0}^{m-u} r_{z,j} \cdot b_m \cdot a_{u-z}^{(k-1)\otimes} \cdot a_{m-u-g}^{(l-k)\otimes} \cdot$$

$$\overline{a_{g+1}} \cdot \sum_{t=0}^{y_{max}\cdot(k-1)} \sum_{x=y_{min}}^{y_{max}} y_t^{(k-1)\otimes} y_x \cdot \overline{y_{L-j-t-x}^{(l-k)\otimes}} \sum_{n=1}^{x} \sum_{d=0}^{b_{max}\cdot F} b_d^{F\otimes}$$

where $F = \lceil \frac{j+n+t}{K} \rceil - 1$. Note that F can be greater than one under case 2.

We derived by means of these expressions the proportional values for $P(W = s \mid C1)$ and $P(W = s \mid C2)$. Let us denote these proportional expressions with $P^*(W = s \mid C1)$ and $P^*(W = s \mid C2)$. Normalizing these values, we obtain

$$P(W = s \mid C1) = \frac{P^*(W = s \mid C1)}{\sum_{s=0} P^*(W = s \mid C1)} \qquad (4.62)$$

$$P(W = s \mid C2) = \frac{P^*(W = s \mid C2)}{\sum_{s=0} P^*(W = s \mid C2)} \qquad (4.63)$$

Finally, we attain the waiting time distribution by

$$P(W = s) = P(W = s \mid C1) \cdot P(C1) + P(W = s \mid C2) \cdot P(C2) \quad (4.64)$$

4.4.5. Numerical Results

Validation of the approach

The method represented for the analysis of the $G^X/G^{[L,K]}/1$-queue is exact within an ϵ-neighborhood. In particular, the quality of the results is affected by the decision on r^y_{max}, the maximum value of the remainder (see section 4.4.1). In this section, we study the accuracy of our approach. For this purpose, we compare simulation results with the results, derived by means of discrete time analysis. We have chosen the configuration in table A.1 in appendix.

We proceed as follows. With simulation, we generate a number of replications (n). For each value i, that the investigated distribution may assume, we compute a confidence interval for the mean probability. As the mean and the standard deviation are unknown, we use $E_i(n)$ and $S_i(n)$, which are estimated based on a sample of n replications. Consequently, we use the t-distribution with $(n-1)$ degrees of freedom at a significance level of α. Finally, it is checked, if the probability value determined by means of the discrete approach lies in the confidence interval. The confidence interval is computed as follows

$$\left(E_i(n) - t_{n-1,1-\alpha/2} \cdot \frac{S_i(n)}{\sqrt{n}}, E_i(n) + t_{n-1,1-\alpha/2} \cdot \frac{S_i(n)}{\sqrt{n}} \right)$$

In this analysis, we generated $n = 8$ replications and used a significance level of $\alpha = 0.01$. The results for the departing batch size, inter-departure time, and the waiting time distributions are displayed in appendix (see tables A.4, A.5, and A.6). In our analysis, we tried different values of r_{max}^y. We derived the distributions with the r_{max}^y-values of 5, 11, 22, and 30. As summarized in the tables, all the probability values calculated for $r_{max}^y \geq 11$ lie in the relevant confidence interval. In contrary, some probability values for $r_{max}^y = 5$ lie out of the computed confidence interval. In general, it applies that an underestimated r_{max}^y does worsen the results, but its overestimation has no significant effect on the quality of the results.

Analysis and optimization of a transport system

In this section, we study a transport process between a sending and a receiving hub station. Specifically, shipments from various destinations arrive at the sending station. Each shipment consists of a number of transport units, which is denoted as the batch size (Y). Moreover, the time between the arrivals of successive shipments is referred to as the inter-arrival time (A). At the sending hub station, transport units are collected and sent to the receiving station. A transport process is triggered, provided that a transport vehicle is available and a given minimum capacity utilization is satisfied. Durations of the one-way and round-trips are denoted by OT and RT, respectively. The distributions of the random variables A, Y, OT, and RT are displayed in table A.3 in appendix.

We assume that the capacity of a transport vehicle is limited to $K = 20$ and the required minimum capacity utilization is 70%. Thus, the parameter L can be determined with the following equation:

$$L = \lceil K \cdot \text{minimum capacity utilization} \rceil \qquad (4.65)$$

In particular, we are motivated in this section by the analysis of the effect of different vehicle dispatching strategies with differing vehicle availabilities on the sojourn time of an arbitrary transport unit as well as on the total system costs. The considered strategies and the associated modeling principles are summarized below.

Clocked Provision (T) Under this strategy, a takt time (T) is given. Every T time units after the last departure, a transport vehicle becomes available. If the number of collected transport units at this time instant satisfies the minimum capacity utilization constraint, a transport process takes place. If not, the vehicle waits for additional shipments until the minimum capacity utilization is fulfilled. So as to model this strategy, we employ the $G^X/G^{[L,K]}/1$-queue. The variables A and Y define the arrival process, whereas the service time has a Dirac distribution with a constant value of T. The number of transport vehicles needed to operate such a transport system is computed later in this section as the part of the cost analysis.

Shuttle Transport In accordance with this strategies, one transport vehicle shuttles between the sending and the receiving station. When the transport vehicle returns to the sending station, the number of collected transport units is checked. If this number is greater than or equal to L, the transport process is initiated. In the contrary case, the vehicle waits for more shipments. Similar to the case with clocked provision, we model the system as the $G^X/G^{[L,K]}/1$-queue. The only difference lies in the service time distribution, which corresponds to the distribution of the round-trip time under the shuttle transport.

Immediate Transport Complying with this strategy, the transport process is triggered, as soon as the constraint on the minimum capacity utilization is fulfilled. For this purpose, we model the system as a batch building process under the capacity interval rule. Arrival process is characterized with the variables A and Y. The necessary number of vehicles to abide by this strategy should be determined in advance. This will be explained later in this section.

Vehicle dispatching strategy	Model	Input	Abbreviation
clocked provision	$G^X/G^{[L,K]}/1$-queue	a,y, t $B = T$ L,K	CP T
shuttle transport	$G^X/G^{[L,K]}/1$-queue	a,y b: rt_i L,K	ST
immediate transport	Capacity interval rule	a,y L,K	IT

Table 4.4.: Summary of vehicle dispatching strategies

These strategies are summarized in table 4.4. We investigated in total 20 different scenarios. The first 18 scenarios correspond to the strategy "Clocked Provision" with $T : 10, 11, \cdots, 27$. The reason, why the scenarios with takt time greater than or equal to 28 are not studied, is that the mean number of incoming units during the mean service time is greater than the limit K. Specifically, following condition for the stability is not fulfilled anymore.

$$\frac{E(B) \cdot E(Y)}{E(A)} < K \qquad (4.66)$$

The last two scenarios, we consider, are the shuttle transport and the immediate transport.

Analysis of the sojourn time for different design principles

We study here the effect of vehicle availability on the sojourn time. To serve this purpose, we firstly compute the expected sojourn time for each scenario as the sum of the mean waiting time and the mean one-way trip time. The waiting time arises due to two reasons either the minimum capacity utilization is not fulfilled or no vehicle is available. As the vehicle availability, we define the percentage of the time that at least one vehicle is available for the transport process. We compute it as the expected value of the idle time divided by the expected value of the departure time:

$$\text{Availability} = \frac{E(IT)}{E(DOUT)} \tag{4.67}$$

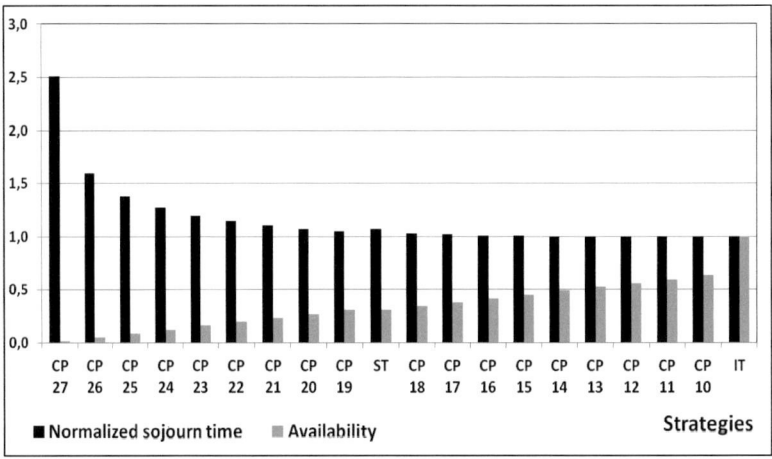

Figure 4.16.: Analysis of the sojourn time with regard to the vehicle availability

In figure 4.16, vehicle availability and the normalized sojourn time for different strategies are depicted. Obvious from the figure is that the sojourn time is pretty stable for a large range of availability values. Indeed, the sojourn time increases drastically, only when the vehicle availability is extremely low. This complies with the phenomenon that the sojourn time or rather the waiting time increases exponentially with increasing utilization of the system. So very low values of vehicle availability should be avoided. Provided that the vehicle availability is not extremely low, it is possible to attain an adequate sojourn time with a modest level of vehicle availability.

Cost analysis of different design principles

In this section, we conduct a cost analysis to identify the most cost efficient strategy. In order to formulate the cost function, we distinguish between two cost categories: transport and inventory costs. Furthermore, each category is subdivided into fixed and variable costs. Assuming an observation period of T_{obs} time units, we derive different types of costs as follows.

Fixed Transport Costs: the number of necessary transport vehicles ($n_{vehicle}$) to run the given strategy is estimated considering the worst case scenario, in which the round-trip time takes its maximum value and the inter-departure time assumes its minimum value. Assuming that the fixed costs per transport vehicle assigned to the observation time period T_{obs} correspond to $C_{vehicle}$ monetary units, we derive the fixed transport costs

$$C_{FTrans} = n_{vehicle} \cdot C_{vehicle} \qquad where \quad n_{vehicle} = \lceil \frac{rt_{max}}{d_{out,min}} \rceil$$

Variable Transport Costs: the average number of transport processes executed in the observation period (n_{trans}) is estimated based on the expected value of the departure time ($E(D_{out})$). Subsequently, we multiply this term with the cost per transport process (C_{trans}) and assess the variable transport costs as given below.

$$C_{VTrans} = n_{trans} \cdot C_{trans} \qquad where \quad n_{trans} = \frac{T_{obs}}{E(D_{out})}$$

Fixed Inventory Costs: we assume here that the storage depot of the sending station is sized with a safety level of 99%. In order to assess the number of necessary storage places n_{capa}, we derive the distribution

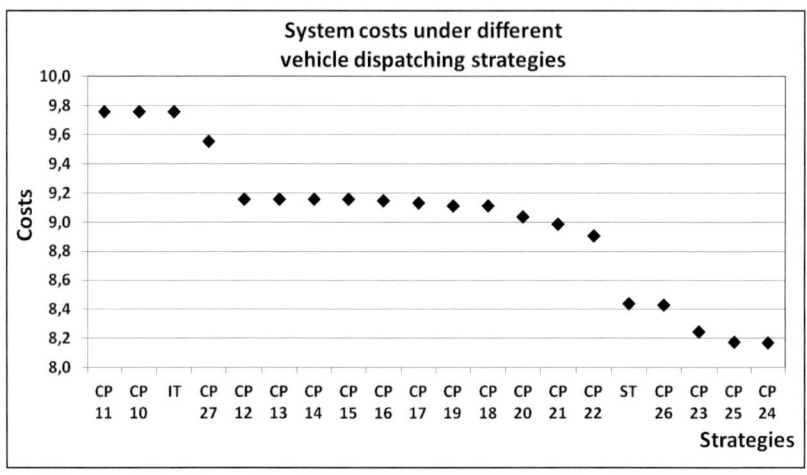

Figure 4.17.: Cost analysis for different scenarios

of the collected number of units immediately before a transport process (y_{col}). We define another variable Y_{stor} as the number of transport units, which were allocated a storage place. In order to derive the distribution of this variable, we limit the distribution y_{col} to n_{capa} (see equation 2.13). Thereafter, we determine the minimum value of n_{capa}, so that the following condition is fulfilled:

$$y_{stor} = \Pi^{n_{capa}}[y_{col}] \qquad \frac{E(Y_{stor})}{E(Y_{col})} \geq 0.99$$

Finally, we attain the fixed inventory costs assigned to the observation time period:

$$C_{FInv} = n_{capa} \cdot C_{capa}$$

where C_{capa} stands for the fixed costs of a storage place allocated to the observation period.

Variable Inventory Costs: the average number of transport units in the system (n_{inv}) is derived in accordance with Little's Law. Based upon n_{inv}, we determine the variable inventory costs.

$$C_{VInv} = n_{inv} \cdot C_{inv} \qquad where \quad n_{inv} = \frac{E(Y)}{E(A)} \cdot E(L)$$

Here, C_{inv}, $E(Y)$, and $E(A)$ correspond to the variable inventory costs (e.g. capital commitment costs) per transport unit in observation time period, the mean values of the batch size and the inter-arrival time, respectively. E(L) stands for the mean sojourn time.

Using the system figures in table A.2 in appendix, we computed the system costs for each scenario. Figure 4.17 summarizes the system costs (in thousand hundred monetary units) for each scenario. The figure shows that the clocked provision strategy with a takt time of 24 time units minimizes the system costs.

5. Vehicle Consolidation

Vehicle consolidation is a spatial consolidation strategy, characterized by employing the same vehicle to serve a couple of receiving and/or shipping points successively. In this chapter, we analyze the milkrun concept, in which vehicle consolidation is applied.

The milkrun concept, introduced by Toyota, is an increasingly important operating mode for the material supply both within and between manufacturing organizations. The idea behind the milkrun concept is to employ the same vehicle(s) to collect deliveries from a number of suppliers and deliver them to a number of customers in a fixed round tour. In a milkrun system, vehicles depart from an outgoing station complying with a given time schedule and follow the routes designed previously. Upon the completion of the route, the vehicle returns back to the initial station. In order to supply the customers with products reliably, it is just as important to supply the suppliers with empty containers. Therefore, the return transport is mostly integrated in the concept.

Due to the consolidation of transport quantities of multiple companies, the milkrun concept enables high-frequency transports while assuring better utilization of transport capacities. Permitting high-frequency deliveries in small batches, the milkrun concept is a crucial component of the JIT production systems and yields a significant reduction in inventory levels. Moreover, the standardization of the routes and the time schedule increases the transparency of the transport process. In this way, it is possible to reduce administrative costs.

Milkrun systems must be designed attentively. Besides the design of the routes, the performance of a milkrun system is closely related to the time schedule. Schedules with smaller time intervals between two successive milkrun tours result in lower inventory levels, while increasing the transport costs. Moreover, the milkrun concept causes a dependency between the visited stations, e.g. a delay at a given station may shift the time schedule for the successive stations and result in fluctuations in transport processes. These fluctuations must be quantified to assure a certain safety level for the on-time order fulfillment. Therefore, analytical models are needed, with which it is possible to study the system dynamics and quantify the performance of the whole system. In this chapter, we suggest the analysis of milkrun systems by means of discrete polling models and propose two models for the analysis of milkrun systems. Referring to the basic polling model presented in section 3.2, the vehicles are the servers, and the visited stations are the queues.

The first model, we introduce, is the takted milkrun system, in which a new tour is initiated complying with a given takt time. The takted milkrun concept is applied very often for vehicle-based internal or external transport processes. A typical example is the supply of assembly stations in automotive industry, where assembly stations store a limited quantity of required components. In fixed time intervals, the components used are replaced by milkrun tours. Therefore, assembly stations must just maintain the quantity needed to cover the consumption in takt time and a safety stock to hedge against fluctuations in supply processes. In this way, space requirement of assembly stations is reduced. Similarly, manufacturing organizations may reduce the capacity requirement of their warehouses by picking up the material from their suppliers or delivering goods to the customers in takted milkrun tours. The second model, we analyze, is the shuttle milkrun system, in which a vehicle shuttles between a number of stations. Many intra-logistics systems, which involve the implementation of line deliveries (e.g. AGV systems), can be modeled as shuttle milkrun systems.

In order to assure a broad range of application area for both models, we avoid in our models restricting assumptions as much as possible. Therefore, we allow a bidirectional material flow by considering not only the flow of goods, but also the flow of e.g. empty containers. This is achieved by a transport matrix, with which it is possible to model arbitrary transport relations between the visited stations. Consequently, a station can be a pure customer or a pure supplier or a combination of both. Thus, the models are applicable to:

- collection tours: transport units are picked up from a number of stations and delivered to exactly one station,

- delivery tours: transport units are picked up at exactly one station and distributed to the other stations,

- pick-up and delivery tours: a station may deliver transport units to an arbitrary number of stations and may receive transport units from an arbitrary number of stations.

As the service discipline for the vehicles, we have chosen an hybrid regime, the gated limited policy. In accordance with the gated limited service policy, only a limited number of transport units are picked up at each station. Besides, the transport units arriving to the shipping area of the given station after the vehicle started the loading process are not loaded in the current tour. For the unloading process, there is no limit; all the units that must be unloaded at a specific station, are unloaded in the given tour. Moreover, the correlation between the handling time (i.e. loading or unloading time) and the quantity to be handled is taken into consideration. So the time, that the vehicle spends at an arbitrary station, is proportional to the quantity loaded or unloaded at this station in the developed models.

This chapter is organized as follows. In section 5.1, we study the takted milkrun systems. For the takted milkrun systems, we firstly present an approximative iterative algorithm. Later on, we introduce an improvement algorithm, which can be used to improve the results of the basic algorithm. In section 5.1, we investigate the shuttle milkrun systems. The algorithm introduced is approximative as well. For these systems, we compute the distributions of the queue states, tour time, cycle times, and the waiting times.

5.1. Takted Milkrun Systems

In the takted milkrun system, vehicles depart from the source (station 0) in fixed time intervals. We refer to the fixed time interval between the successive departures from station 0 as the takt time (T). Every T time units, a new tour starts, in which a fixed number of stations (N) are visited. The course of each tour involves the following steps:

1. load the transport units at station 0 (if the station delivers some transport units),

2. drive to the next station,

3. unload the transport units, of which destination is the given station (if there is any),

4. load the transport units at the given station (if the station delivers some transport units),

5. repeat steps 2-4 till station (N-1),

6. drive back to station 0,

7. unload the transport units, of which destination is station 0 (if there is any),

8. the vehicle starts waiting in a queue of vehicles till it is removed from the queue for a new tour.

Note that the vehicle on tour starts firstly the unloading process and then the loading process at an arbitrary station. Moreover, we assume in our analysis that there may be more than one vehicle in the system, in order to manage the takt time. Upon the completion of a tour, the vehicle starts waiting in a queue of vehicles, until it is removed from the queue for a new tour. While waiting in the queue, the vehicle may not be empty. Assume, for instance, two stations, one of which is visited as the fourth and the other one as the sixth station in a tour. If products or containers are sent from the sixth station to the fourth, then this is possible earliest in the next tour. After collecting the quantity at the sixth station, the vehicle completes the tour and returns back to station 0. At station 0, the vehicle waits with the transferred quantity till it is removed from the queue for a new tour. We assume here, that the quantities loaded on a vehicle cannot be transferred to another vehicle that stands in front of the given vehicle in the queue. Depending on the number of vehicles in the queue, the vehicle will be removed from the queue in the next tours and then it can deliver the quantity to the fourth station.

We assume further that a vehicle cannot overtake another vehicle, which started the tour earlier. In this situation, the vehicle slows down and starts following the previous vehicle. Due to this assumption, our analysis applies best for the cases, in which the possibility of an overtake is pretty small. This is usually the case in industrial practice, where the takt time is at least comparable with the maximum driving time.

5.1.1. Queuing System

In this section, we investigate a takted milkrun system with N stations assuming a bidirectional material flow. To account for this assumption, each station possesses a loading and an unloading station with infinite buffers. The transferred quantities between different stations are determined based on a given transport matrix, which summarizes, how much percentage of the units loaded at an arbitrary station j is delivered to station i. So an arbitrary transport unit loaded at station j is transferred to station i with a probability $p^{j,i}$.

In our model, transport units arrive at an arbitrary station i in random time intervals of length A^i and in random batch sizes of Y^i. If the station does not deliver any transport units to other stations, then there is no arrival process at this station. The variable switch-over time S^i refers to the driving time from an arbitrary station i to the next station. We assume that the loading and the unloading time for a transport unit have the same distribution. The service time B represents the loading or the unloading time of a single unit. These random variables are discrete and iid. Complying with the gated limited policy, a limit K^i is defined for each station, which stands for the maximum number of transport units, that can be loaded at station i in each tour. In this analysis, we employ the following variables, some of which are illustrated in figure 5.1.

T takt time,

K^i constant limit for the quantity picked up (loaded) at station i in a tour,

S^i driving time between station i and the next station,

A^i inter-arrival time at station i,

Y^i incoming batch size at station i,

R^i residual inter-arrival time at station i,

B loading/unloading time for a transport unit,

$p^{j,i}$ probability, that an arbitrary transport unit loaded at station j is delivered to station i,

X^i queue state at station i, number of transport units, that the vehicle sees immediately before the start of a loading process at station i,

$\overline{X}^i$ number of transport units loaded at station i in a tour,

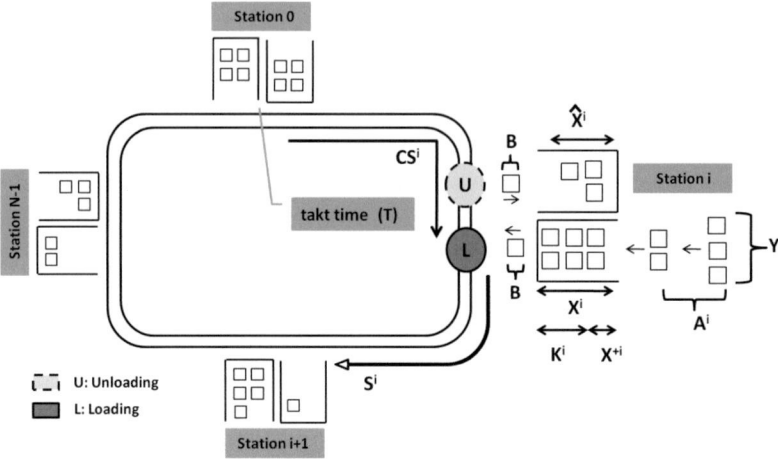

Figure 5.1.: Illustration of a takted milkrun system and the related variables employed in the analysis

X^{+i} number of transport units beyond the limit at the start of a loading process at station i,

$\widehat{X}^{j,i}$ number of transport units delivered (transferred) from station j to station i,

$\widehat{X}^{i}$ number of transport units unloaded at station i,

CS^{i} cycle segment at station i, total time from beginning of a tour until beginning of loading process at station i,

C^{i} cycle time at station i, time interval between two successive loading processes at station i,

G^{i} number of transport units arrived in a cycle at station i,

TT tour time, duration of a tour,

W^{i} waiting time of a transport unit at station i

WL^{i} workload till station i, sum of picked and delivered quantities from the beginning of a tour until the beginning of the loading process at station i,

TWL total workload in a tour, sum of picked and delivered quantities in a tour.

A queuing station i is stable, if the mean number of units arriving during the mean cycle time $E(C^i)$ is less than the limit K^i. As the mean cycle time at each station can be replaced by the takt time (T), following condition is fulfilled for stable systems:

$$\frac{T \cdot E(Y^i)}{E(A^i)} < K^i \qquad \forall i = 0, 1, \cdots, N-1 \tag{5.1}$$

Moreover, the takt time must be long enough, so that the possibility of an overtake between the vehicles is insignificant. To check this assumption, one may compute the difference between the maximum and the minimum tour time and verify that the takt time is at least comparable with this figure. Thus, the model is **not** applicable to cases where

$$T << (\sum_{i=0}^{N-1} s^i_{max} + \sum_{i=0}^{N-1} 2 \cdot E(B) \cdot K^i) - \sum_{i=0}^{N-1} s^i_{min}$$

Given the inputs T, K^i, $p^{j,i}$, and the distributions a^i, y^i, s^i, and b, we derive approximately the steady state distributions of the cycle time (c^i), queue states (x^i), tour time (tt), and the waiting time of an arbitrary transport unit (w^i) for stable systems.

5.1.2. Iterative Algorithm

Similar to the approach presented in Dittmann and Hübner (1993), we propose here an iterative algorithm to approximate the cycle times (C^i), queue states (X^i) as well as the tour time (TT). The steps of the algorithm are summarized as follows:

1) Initialize the queue states (X^i), e.g. by setting the system size to zero,

2) Calculate the cycle segment (CS^i) from (X^{i-1}) (see section 5.1.3) and (X^i) from (CS^i) for each station (see section 5.1.4). Repeat the step for $i = 0, 1, 2, \cdots, (N-1)$,

3) Calculate the tour time (TT) (see section 5.1.5),

4) Repeat steps 2)-3) until a convergence criterion is fulfilled,

5) Compute the cycle times (C^i) for $i = 0, 1, 2, \cdots, N-1$ (see section 5.1.6).

Complying with this algorithm, the distributions of the cycle segment, queue states, and the tour time are updated in each iteration step based on their distributions from the previous iteration step. The algorithm terminates when the given convergence criterion is satisfied. A convergence criterion can be, for instance, the absolute difference between the mean tour duration from the actual iteration step and that from the previous iteration step (e.g. we use in our analysis $|E(TT^{n+1}) - E(TT^n)| \le 0.0001$ as the convergence criterion). When the algorithm is aborted, the cycle time distributions can be computed.

Initialization

We assume, as in the previous analysis, the zero-fold convolution of the input variables are Dirac distributed with a constant value of zero:

$$a_0^{i,0\otimes} = y_0^{i,0\otimes} = b_0^{0\otimes} = 1 \qquad \forall i = 0, 1, \cdots, (N-1)$$

Before the execution of the iterative algorithm, we approximate the residual inter-arrival time at each station i by that of a renewal process. As the observation of the residual inter-arrival time takes place immediately after discrete time instants, we approximate it as the residual lifetime of a renewal process observed immediately after the event occurrence. Hence, it yields

$$r_s^i = \frac{1}{E(A^i)}(1 - \sum_{v=0}^{s-1} a_v^i) \qquad \forall s = 1, \cdots, a_{max}^i \tag{5.2}$$

Concluding, we set the initial queue states to zero:

$$x_0^i = 1 \qquad \forall i = 0, 1, \cdots, (N-1)$$

5.1.3. Cycle Segment Distributions

The cycle segment (CS^i) defines the time interval between the beginning of a tour and the beginning of the loading process at station i. The

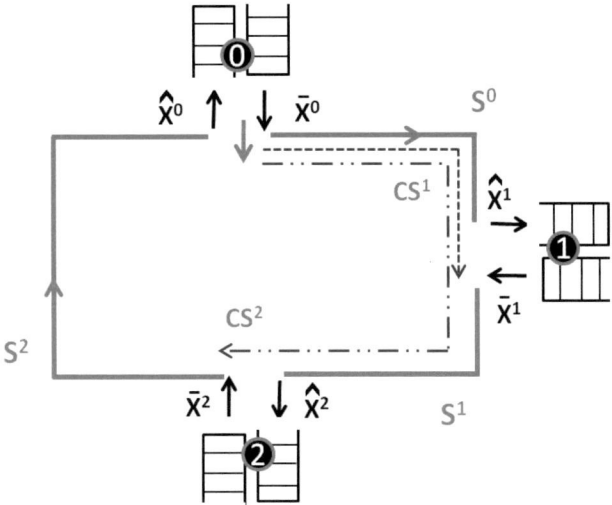

Figure 5.2.: Illustration of the cycle segment in a takted milkrun system

figure 5.2 illustrates the variable cycle segment for a system with three stations. Since the tour starts with loading process at station 0, the cycle segment is always zero for station 0:

$$cs_0^0 = 1 \tag{5.3}$$

As illustrated in the figure, the cycle segment CS^1 consists of the loading time at station 0, unloading time at station 1, and the driving time between station 0 and station 1. Having computed CS^1, one can compute CS^2 as the sum of CS^1, loading time at station 1, unloading time at station 2, and the driving time between station 1 and station 2. Thus, it yields mathematically:

$$
\begin{aligned}
CS^0 &= 0 \\
CS^1 &= CS^0 + (\overline{X}^0 + \widehat{X}^1) \cdot B + S^0 \\
CS^2 &= CS^1 + (\overline{X}^1 + \widehat{X}^2) \cdot B + S^1
\end{aligned}
\tag{5.4}
$$

Generalizing this, we attain the cycle segment at an arbitrary station i:

$$CS^i = CS^{i-1} + (\overline{X}^{i-1} + \widehat{X}^i) \cdot B + S^{i-1} \qquad \forall i = 1, \cdots, (N-1) \quad (5.5)$$

The variables CS^{i-1} and $\overline{X}^{i-1}$ in equation 5.5 are dependent on each other. Verbally, if it takes e.g. longer in a tour till the vehicle can start with loading at station $(i-1)$, then there is a higher probability to see an above-average queue length (X^{i-1}) at this station. Thus, the picked quantity at this station $(\overline{X}^{i-1})$ will have an above-average value. Considering this dependency, we compute the cycle segment distribution at station i as in equation 5.7. Note that we omit here some other kinds of dependencies, which we will explain in further detail with the introduction of the improvement algorithm (see section 5.1.8).

for $i \geq 1$,

$$P(CS^i = m + d + e) = \sum_{m=0}^{cs_{max}^{i-1}} \sum_{n=0}^{K^{i-1}} cs_m^{i-1} \cdot (\overline{x}_n^{i-1} \mid CS^{i-1} = m) \quad (5.6)$$

$$\sum_{z=0}^{\widehat{x}_{max}^i} \sum_{d=0}^{(n+z) \cdot b_{max}} \sum_{e=s_{min}^{i-1}}^{s_{max}^{i-1}} \widehat{x}_z^i \cdot b_d^{(n+z)\otimes} \cdot s_e^{i-1}$$

In equation 5.7, the cycle segment and the quantity loaded at station $(i-1)$ (conditioned on the cycle segment) appear. We need to know the distributions of these variables given by cs^{i-1} and $(\overline{x}^{i-1} \mid CS^{i-1} = m)$, respectively. In accordance with the iterative algorithm, we firstly calculate the cycle segment (CS^{i-1}) for station $(i-1)$ and then we compute the queue state X^{i-1}. In order to derive the queue state, we firstly compute the queue state conditioned on the cycle time $(X^{i-1} = n \mid CS^{i-1} = m)$, as explained in the next section. Thus, when we compute the distribution of the cycle segment for the next station (cs^i), the cycle segment distribution at the previous station (cs^{i-1}) and the conditional distribution $(x^{i-1} \mid CS^{i-1} = m)$ are already known. The distribution $(\overline{x}^{i-1} \mid CS^{i-1} = m)$ corresponds to the distribution $(x^{i-1} \mid CS^{i-1} = m)$ with an upper bound K^{i-1}. With the operator introduced in equation 2.13, we attain:

$$(\overline{x}^{i-1} \mid CS^{i-1} = m) = \Pi^{K^{i-1}}[(x^{i-1} \mid CS^{i-1} = m)] \tag{5.7}$$

Another expression that appears in the equation is the quantity delivered to station i in a tour, which is defined by the variable $\widehat{X}^i$. The delivered quantity at station i is the sum of all quantities delivered from other stations to station i. Hence,

$$\widehat{X}^i = \sum_{j=0}^{N-1,j \neq i} \widehat{X}^{j,i} \tag{5.8}$$

In order to compute $\widehat{X}^i$, we proceed to derive the distribution of the transferred quantity from an arbitrary station j to station i.
for $m = 0, 1, \cdots, K^j$,

$$P(\widehat{X}^{j,i} = m) = \sum_{n=m}^{K^j} P(\overline{X}^j = n) \binom{n}{m} (p^{j,i})^m (1 - p^{j,i})^{n-m} \tag{5.9}$$

In equation 5.9, we assume that the number of units loaded at station j equals an arbitrary value n. The probability that the transferred quantity from station j to station i is $m \leq n$, corresponds to the binomial probability of having m successes in n trials. Note that the maximum value of n is K^j. Neglecting the dependencies between the transferred quantities from different stations, we attain the distribution $\widehat{x}^i$:
for $j \neq i$,

$$\widehat{x}^i = \widehat{x}^{0,i} \otimes \widehat{x}^{1,i} \otimes \cdots \otimes \widehat{x}^{j,i} \otimes \cdots \otimes \widehat{x}^{(N-1),i} \tag{5.10}$$

5.1.4. Queue State Distributions

In this section, we derive the queue state X^i at an arbitrary station i, which is the number of transport units that the vehicle sees immediately before the start of the loading process at the given station. For this purpose, we firstly derive the queue state conditioned on the cycle segment $(X^i \mid CS^i = m)$, which is also needed to derive the cycle segment for the next queue (see section 5.1.3).

The queue state X^i at station i is the sum of transport units beyond the limit at the start of the previous loading process (X^{+i}) and the number of transport units arrived within the cycle (G^i). In figure 5.3, these variables are illustrated for an example. Obviously, the variable (X^{+i}) does not depend on the length of the cycle segment realized in the actual tour. But the variable G^i depends on the cycle time (C^i), which stands for the time interval between two successive loading processes at station i. As will be explained later in this section, the cycle time depends on the cycle segment in the current tour. Hence, the variable G^i depends also on the cycle segment. We obtain:

$$(X^i \mid CS^i = m) = X^{+i} + (G^i \mid CS^i = m) \tag{5.11}$$

Firstly, we derive the distribution of X^{+i}. The number of transport units loaded at station i in the previous tour is $\min(X^i, K^i)$. Therefore, it yields:

$$X^{+i} = X^i - \min(X^i, K^i) = \max(0, X^i - K^i) \tag{5.12}$$

The difference between a random variable and a constant value corresponds to a negative shift of K^i units in the distribution of the random variable. Thus, we use the shift operator, introduced in equation 2.14, to compute the auxiliary distribution:

$$\tilde{x}^{+i} = \Delta_{-K^i}[x^i] \tag{5.13}$$

The auxiliary distribution $\tilde{x}^{+i}$ may get negative values, thus we obtain the distribution x^{+i} by lower bounding it to zero (see equation 2.13):

$$x^{+i} = \Pi_0[\tilde{x}^{+i}] \tag{5.14}$$

As the next, we compute the distribution $(g^i \mid CS^i)$ as follows:

$$P(G^i = k \mid CS^i = m) = \sum_{t=0}^{c_{max}^i} (g_k^i \mid C^i = t) \cdot (c_t^i \mid CS^i = m) \tag{5.15}$$

where $P(G^i = k \mid C^i = t)$ is the probability that k transport units arrive at station i, given a cycle time of t time units, and is equal to:

$$P(G^i = k \mid C^i = t) = \sum_{s=1}^{a_{max}^i} \sum_{l=1}^{\left\lceil \frac{t}{a_{min}^i} \right\rceil} \sum_{g=0}^{t-s} r_s^i \cdot a_{t-s-g}^{i,(l-1)\otimes} \cdot \overline{a_{g+1}^i} \cdot y_k^{i,l\otimes} \tag{5.16}$$

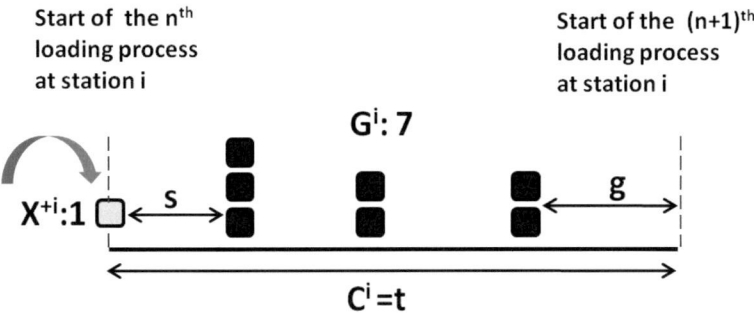

Figure 5.3.: Derivation of the queue states in a takted milkrun system

In the previous equation, we assume that l batches arrive in t time units and have a total size of k transport units. And with the sum of steady state probabilities, we get

$$P(G^i = 0 \mid C^i = t) = 1 - \sum_{k=1}^{\infty} P(G^i = k \mid C^i = t) \tag{5.17}$$

Subsequently, we need to compute the probability $P(C^i = n \mid CS^i = m)$, which is also needed in equation 5.15. To do so, we firstly study the relationship between the cycle segment and the cycle time. As explained beforehand, the cycle time at station i is defined to be the time between the start of the loading process in the previous tour and that in the actual tour. Neglecting the dependencies between the courses of two successive tours, we assume that independent observations of the cycle segment CS^i are realized in two successive tours. As illustrated in figure 5.4, cycle time realized in the $(n + 1)^{th}$ tour is $C^{i,n+1} = CS^{i,n+1} + T - CS^{i,n}$

Now, we can derive the the cycle time distribution dependent on the length of the cycle segment in the current tour. Given that the cycle segment till station i was n time units in the previous tour and m time units in the actual tour, then the cycle time is $(m + T - n)$ time units (see figure 5.4). Hence, we get:

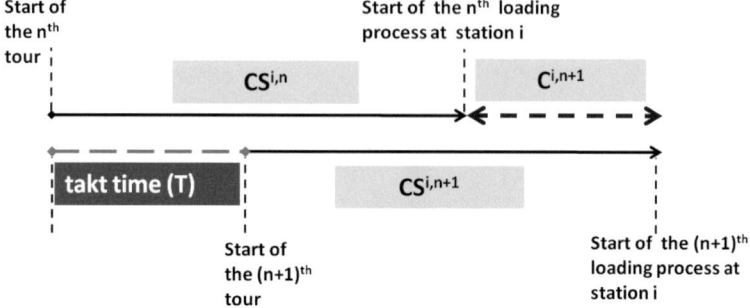

Figure 5.4.: Illustration of the relationship between the cycle time and the cycle segment in a takted milkrun system

for $m + T > n$,

$$P(C^i = m + T - n \mid CS^i = m) = P(CS^i = n) \tag{5.18}$$

Recall that the possibility of an overtake is excluded in our analysis. We assume, that the vehicle, that started the tour later, slows down and follows the previous vehicle. Thus,

$$P(C^i = 0 \mid CS^i = m) = \sum_{k=m+T}^{cs^i_{max}} P(CS^i = k) \tag{5.19}$$

Note that the cycle segment for station 0 is always equal to zero. Consequently, the cycle time at station 0 is always equal to the takt time.

Having obtained the distributions x^{+i} and $(g^i \mid CS^i = m)$, we convolute them to attain the queue state distribution, given the cycle segment.

$$(x^i \mid CS^i = m) = x^{+i} \otimes (g^i \mid CS^i = m) \tag{5.20}$$

Eventually, the queue state is derived with the law of total probability. Therefore, we obtain

$$P(X^i = k) = \sum_{m=0}^{cs^i_{max}} P(X^i = k \mid CS^i = m) \cdot P(CS^i = m) \tag{5.21}$$

5.1.5. Tour Time Distribution

We computed the cycle segments and the queue states for all stations from 0 to $(N-1)$. Subsequently, the tour time $(TT = CS^{N-1} + (\overline{X}^{N-1} + \widehat{X}^0) \cdot B + S^{N-1})$ can be calculated in a similar fashion like in equation 5.7:

$$P(TT = m + d + e) = \qquad (5.22)$$

$$\sum_{m=0}^{cs_{max}^{N-1}} \sum_{n=0}^{K^{N-1}} cs_m^{N-1} \cdot (\overline{x}_n^{N-1} \mid CS^{N-1} = m)$$

$$\sum_{z=0}^{\widehat{x}_{max}^0} \sum_{d=0}^{(n+z) \cdot b_{max}} \sum_{e=s_{min}^{N-1}}^{s_{max}^{N-1}} \widehat{x}_z^0 \cdot b_d^{(n+z)\otimes} \cdot s_e^{N-1}$$

5.1.6. Cycle Time Distributions

After the iterative algorithm is aborted, we can compute the cycle time distribution as follows:

$$P(C^i = n) = \sum_{m=0}^{c_{max}^i} P(C^i = n \mid CS^i = m) \cdot P(CS^i = m) \qquad (5.23)$$

The expressions $P(CS^i = m)$ and $P(C^i = n \mid CS^i = m)$ are computed in the last iteration step before the algorithm terminates (see section 5.1.3 and section 5.1.4, respectively).

5.1.7. Waiting Time Distributions

Having computed the cycle time distributions, we investigate the waiting time of an arbitrary transport unit at station i. Due to the gated limited service regime, each transport unit at station i has to wait till the beginning of the next loading process. Assuming that the unit arrives t time units after the start of the current cycle of length $m \geq t$, then the unit waits $(m - t)$ time units until the next loading process starts. Moreover, the unit has to wait until all the units, that arrived earlier, are loaded. Those transport units include the units arrived in the previous cycle but could not be loaded due to limit (X^{+i}) and the units arrived within the current cycle ahead of the observed unit (δ). The components of the waiting time of an arbitrary transport unit are illustrated in figure 5.5. Once, the number of units in front of the observed unit is known, it is possible to compute their loading time.

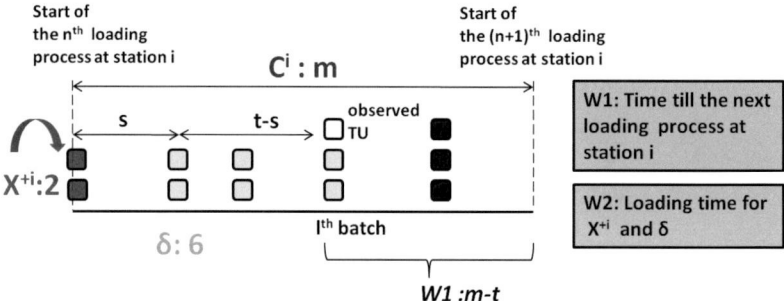

Figure 5.5.: Derivation of the waiting time of an arbitrary transport unit at an arbitrary station i in a takted milkrun system

Let us investigate the case displayed in figure 5.5. In this example, $(X^{+i} = 2)$ and $(\delta = 6)$, so there are 8 transport units that should be handled before the observed unit can be loaded on the vehicle. If $(K^i = 5)$, then the observed unit has to wait a complete cycle time and the loading time for three transport units. Based on this logic, we firstly compute the waiting time of an arbitrary unit, given that the unit arrives within a cycle interval of m time units. This probability is proportional to:

$$P(W^i\text{=m-t+z+e} \mid C^i = m) \approx \qquad (5.24)$$

$$\sum_{n=0}^{x_{max}^{+i}} \sum_{s=1}^{a_{max}^i} \sum_{t=s}^{m} \sum_{l=1}^{\lceil \frac{t}{a_{min}^i} \rceil} \sum_{x=(l-1)\cdot y_{min}^i}^{(l-1)\cdot y_{max}^i} \sum_{j=y_{min}^i}^{y_{max}^i} \sum_{h=0}^{j-1} \sum_{z=0}^{f\cdot c_{max}^i} \sum_{e=0}^{I\cdot b_{max}} x_n^{+i} \cdot$$

$$r_s^i \cdot a_{t-s}^{i,(l-1)\otimes} \cdot y_x^{i,(l-1)\otimes} \cdot y_j^i \cdot j \cdot c_z^{i,F\otimes} \cdot b_e^{I\otimes}$$

where

$$F = \lfloor \frac{\text{n+x+h}}{K^i} \rfloor \qquad \text{and} \qquad I = \text{n+x+h} - F \cdot K^i$$

In equation 5.25, the number of units that stand before the observed units is given by $(n + x + h)$. If $(n + x + h)$ is greater than the limit K^i, then the unit has to wait a multiple of cycle times additionally. The number of complete cycle times, that the unit has to wait additionally, is computed by the variable F and the variable I is used to compute the additional loading time, that the observed unit has to wait. We denote the proportional values for $P(W^i = s \mid C^i = m)$ with $P(\tilde{W}^i = s \mid C^i = m)$. Normalizing these proportional values, it yields:

$$P(W^i = s \mid C^i = m) = \frac{P(\tilde{W}^i = s \mid C^i = m)}{\sum_{k=0}^{\infty} P(\tilde{W}^i = k \mid C^i = m)} \tag{5.25}$$

We have computed the waiting time distribution given that the arbitrary unit arrives at station i within a cycle of m time units. With the law of total probability, we compute the waiting time:

$$P(W^i = s) = \sum_{m=1}^{\infty} P(W^i = s \mid C^i = m) \cdot c_m^i \cdot \frac{m}{E[C^i]} \tag{5.26}$$

The equation 5.26 is weighted by the cycle length (m) in relation to $E[C^i]$, since the probability that the unit arrives within a cycle of m time units increases with increasing m.

5.1.8. Improvement Algorithm

The basic algorithm, we presented in the previous section, takes account of the dependency between the cycle segment (CS^i) and the quantity picked up $(\overline{X}^i)$ at an arbitrary station. This dependency is based on the fact, that e.g. more transport units arrive at an arbitrary station, if it takes longer before the vehicle starts with loading at the given station. Therefore, it is more probable that more transport units are loaded at this station. However, the algorithm omits the dependency between the picked and delivered quantities in a tour.

To illustrate this dependency, let us consider a takted milkrun system with 4 stations. Assuming that the quantity picked up at station 2 was five transport units, thus, $\overline{X}^2 = 5$, the transferred quantities from station 2 to station 3 ($\widehat{X}^{2,3}$) and from station 2 to station 0 ($\widehat{X}^{2,0}$) cannot be more than five transport units in this tour. Because stations 3 and 0 are visited after the vehicle has picked up the transport units at station 2. Moreover, if we know that $\widehat{X}^{2,3}$ was equal to three units, $\widehat{X}^{2,0}$ cannot be more than two units. However, the transferred quantity ($\widehat{X}^{2,1}$) can be more than five transport units in this tour. This is due to the fact, that station 1 is visited earlier than station 2 in a tour. Thus, the transferred quantity ($\widehat{X}^{2,1}$) originates from the quantity picked up at station 2 in an earlier tour and is independent of the quantity, that is picked at station 2 in this tour.

In the case that the quantity picked up at an arbitrary station i has exactly one destination j, thus, $\widehat{X}^{i,j} = \overline{X}^i$ and $p^{i,j} = 1$, the dependency between the picked quantities and the delivered quantities becomes more significant. Thus, the results of the basic algorithm may show some deviations from the actual results. Especially, for the systems, in which each station delivers to not more than one station, thus $p^{i,j} = \{0,1\}$ for all transport relations, the distributions of the tour time and the cycle times from the iterative algorithm may show relatively larger deviations. In contrary, the quality of the algorithm for the queue state and waiting time distributions is still very good (see numerical results in section 5.1.9). For systems with $p^{i,j} = \{0,1\}$ for all transport relations, we propose an improvement algorithm, which uses the queue state distributions from the iterative algorithm, and improves the cycle time and tour time distributions. Besides, it is possible to improve the queue states and the waiting time distributions based on the improved cycle time distributions. However, as the basic algorithm performs already well regarding the waiting time and queue state distributions, the improvement algorithm yields only a slight improvement for these performance measures.

With the improvement algorithm, we present here, the dependencies, mentioned above, are considered in the following ways. Firstly, we consider the dependency between the quantity loaded at an arbitrary station and the progress of the tour till the loading process at the station.

In the iterative algorithm, this is achieved by considering the dependency between the cycle segment (CS^i) and the picked quantity $(\overline{X}^i)$. As the length of the cycle segment is correlated with the number of transport units picked or delivered between the beginning of the tour and the beginning of the loading process at the given station, we introduce here the variable workload WL^i, which stands for this number. The relationship between the cycle segment and the workload is given by the following formula:

$$CS^i = (WL^i) \cdot B + S^0 + S^1 + \cdots + S^{i-1} \qquad (5.27)$$

The service time (B) and the driving times (S^j) are independent of the workload. That's why, we can also consider directly the dependency between the workload (WL^i) and the picked quantity $(\overline{X}^i)$, instead of the dependency between the cycle segment and the picked quantity. The improvement algorithm is based on this fact.

Secondly, the possible dependency between the picked and delivered quantities in the current tour is considered. To do so, we differentiate between dependent and independent delivered quantities. The delivered quantity $\widehat{X}^{i,j}$ is dependent on the picked quantity at station i $(\overline{X}^i)$ and the delivered quantities $(\widehat{X}^{i,i+1}, \cdots, \widehat{X}^{i,j-1})$ in the current tour, if the source station i is visited earlier than the drain station j in a tour, which is determined by the route design. Thus, for $i < j$ or $j = 0$, $\widehat{X}^{i,j}$ is dependent on the course of the actual tour. Otherwise, the quantity $\widehat{X}^{i,j}$ is independent of the course of the actual tour, since it originates from the quantities picked in a previous tour. Obviously, independent quantities $\widehat{X}^{i,j}$ (for $i > j$ or $j \neq 0$) can be correlated with each other. For instance, the quantities $\widehat{X}^{3,1}$ and $\widehat{X}^{3,2}$ are independent of the picked quantity $\overline{X}^3$ in the actual tour. But they are dependent on each other, as they originate from the same picked quantity in an earlier tour. For instance, their sum cannot be greater than K^3. However, we omit the dependencies between the quantities delivered from an earlier tour.

In accordance with the improvement algorithm, we compute the workload WL^i for each station $i = 1, 2, \cdots, (N-1)$ considering possible dependencies in the current tour. The steps of the improvement algorithm are listed below:

1) Take the queue states (X^i) from the iterative algorithm. Calculate $(\overline{X}^i)$ with $\overline{x}^i = \Pi^{K^i}[x^i]$ (see equation 2.13) and $\widehat{X}^{i,j}$ for $i > j$ and $j \neq 0$ with equation 5.9,

2) Calculate the workload (WL^i) from $(\overline{X}^{i-1} \mid WL^{i-1})$,

3) Calculate $(C^i \mid WL^i)$ from WL^i and the picked quantity $(\overline{X}^i \mid WL^i)$ from $(C^i \mid WL^i)$,

4) Calculate the cycle time (C^i) from $(C^i \mid WL^i)$,

5) Repeat the steps 2)-4) for $i = 1, 2, \cdots, N - 1$,

6) Calculate the tour time TT.

Workload distributions

Referring to the example in figure 5.2, we investigate, first of all, the workloads for different stations. The workload, before the loading process at station 0, WL^0, is always equal to zero, as the vehicle starts a tour with the loading process at station 0. Workload till station 1 (WL^1) is, on the other hand, the sum of the quantity picked up at station 0 and the quantity delivered to station 1. Eventually, WL^2 is the sum of WL^1, the quantity picked up at station 1 and the quantity delivered to station 2. Hence;

$$
\begin{aligned}
WL^0 &= 0 \\
WL^1 &= \overline{X}^0 + \widehat{X}^1 \\
WL^2 &= \overline{X}^0 + \widehat{X}^1 + \overline{X}^1 + \widehat{X}^2 \\
&\quad \cdots\cdots\cdots \\
WL^i &= \overline{X}^0 + \widehat{X}^1 + \cdots + \overline{X}^{i-1} + \widehat{X}^i
\end{aligned}
$$

Figure 5.6 displays the contents of the workloads at some stations in further detail. For this purpose, we differentiate between the picked quantities and the delivered quantities in two blocks. In the first block, we display the picked quantities and in the second block the delivered quantities. The border in the second block helps to identify the dependent and the independent delivered quantities. On the left side of the border, the dependent quantities are listed. Let us study the content

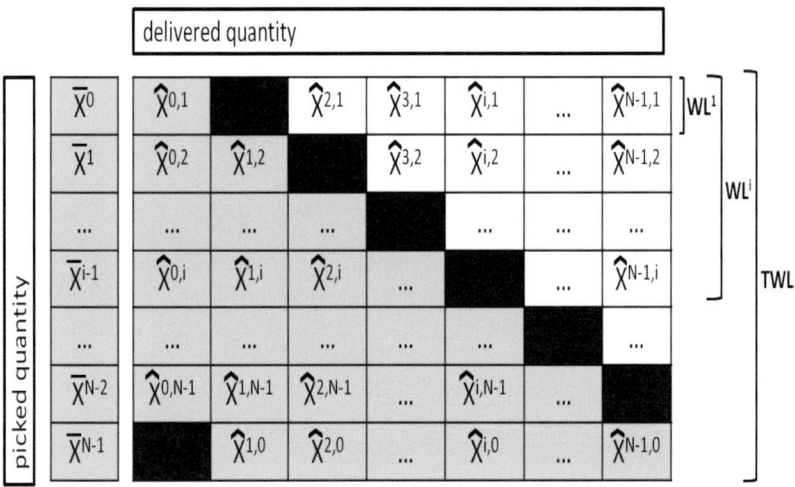

Figure 5.6.: Illustration of the workloads in a takted milkrun system

of the workload at station 1 (WL^1), which includes the picked and the delivered quantities listed in the first line in figure 5.6. Hence;

$$WL^1 = \overline{X}^0 + \underbrace{\widehat{X}^{0,1}}_{\text{dependent on } \overline{x}^0} + \underbrace{\widehat{X}^{2,1} + \widehat{X}^{3,1} + \cdots + \widehat{X}^{N-1,1}}_{\text{independent}} \quad (5.28)$$

For WL^2, we consider the picked and the delivered quantities in the first two lines of the figure, thus:

$$WL^2 = \overline{X}^0 + \underbrace{\widehat{X}^{0,1}}_{\text{dep. on } \overline{x}^0} + \underbrace{\widehat{X}^{2,1} + \widehat{X}^{3,1} + \cdots + \widehat{X}^{N-1,1}}_{\text{indep.}}$$

$$+ \underbrace{\overline{X}^1}_{\text{dep. on } WL^1} + \underbrace{\widehat{X}^{0,2}}_{\text{dep. on } \overline{x}^0 \text{and } \widehat{x}^{0,1}} + \underbrace{\widehat{X}^{1,2}}_{\text{dep. on } \overline{x}^1}$$

$$+ \underbrace{\widehat{X}^{3,2} + \cdots + \widehat{X}^{N-1,2}}_{\text{indep.}} \quad (5.29)$$

Referring to the contents of the first i lines in figure 5.6, we can generalize our analysis to the workload for an arbitrary station i:

$$WL^i = \overline{X}^0 + \underbrace{\widehat{X}^{0,1}}_{\text{dep. on } \overline{X}^0} + \underbrace{\widehat{X}^{2,1} + \widehat{X}^{3,1} + \cdots + \widehat{X}^{N-1,1}}_{\text{indep.}}$$

$$+ \underbrace{\overline{X}^1}_{\text{dep. on } WL^1} + \underbrace{\widehat{X}^{0,2}}_{\text{dep. on } \overline{X}^0 \text{and } \widehat{X}^{0,1}} + \underbrace{\widehat{X}^{1,2}}_{\text{dep. on } \overline{X}^1}$$

$$+ \underbrace{\widehat{X}^{3,2} + \cdots + \widehat{X}^{N-1,2}}_{\text{indep.}} + \cdots + \underbrace{\overline{X}^{i-1}}_{\text{dep. on } WL^{i-1}} +$$

$$+ \underbrace{\widehat{X}^{0,i}}_{\text{dep. on } \overline{X}^0,\widehat{X}^{0,1},\cdots,\widehat{X}^{0,i-1}} + \underbrace{\widehat{X}^{1,i}}_{\text{dep. on } \overline{X}^1,\widehat{X}^{1,2},\cdots,\widehat{X}^{1,i-1}}$$

$$+ \cdots + \underbrace{\widehat{X}^{i-1,i}}_{\text{dep. on } \overline{X}^{i-1}} + \underbrace{\widehat{X}^{i+1,i} + \cdots + \widehat{X}^{N-1,i}}_{\text{indep.}} \tag{5.30}$$

As illustrated in equation 5.30, delivered quantities $\widehat{X}^{i,j}$ (for $i < j$ or $j = 0$) are not only dependent on the quantity picked up at station i ($\overline{X}^i$), but also on the proportion of this quantity that is already delivered to their destinations before reaching station j. For instance, the transferred quantity $\widehat{X}^{0,i}$ is not only dependent on the quantity picked up at station 0, but also on the sum of delivered quantities $\widehat{X}^{0,1}$, $\widehat{X}^{0,2}$,..., $\widehat{X}^{0,i-1}$. Let us assume that $i = 3$ and $N = 5$, in this case the quantity $\widehat{X}^{0,3}$ is not dependent only on $\overline{X}^0$ but also on the quantities $\widehat{X}^{0,1}$ and $\widehat{X}^{0,2}$. Verbally, if the vehicle picks up five transport units at station 0 and delivers two of them to station 1 and two of them to station 2, then $\widehat{X}^{0,3}$ can be at most one in this tour. We define a new variable $U^{i,j}$, to account for the sum of already delivered quantities, that the quantity $\widehat{X}^{i,j}$ is dependent on. Following this notation, $\widehat{X}^{0,3}$ is dependent on $\overline{X}^0$ and $U^{0,3} = \widehat{X}^{0,1} + \widehat{X}^{0,2}$. Now let us investigate the delivered quantity $\widehat{X}^{0,1}$. Clearly, this variable is dependent on $\overline{X}^0$. In this case, no units from this load are delivered to any other stations before reaching station 1. That's why, $U^{0,1}$ is always equal to zero. For $\widehat{X}^{2,0}$, $U^{2,0} = \widehat{X}^{2,3} + \cdots + \widehat{X}^{2,N-1}$. Generalizing these examples, we can attain for $i < j$,

$$U^{i,j} = \begin{cases} 0 & j = i+1 \\ \widehat{X}^{i,i+1} & j = i+2 \\ \widehat{X}^{i,i+1} + \cdots + \widehat{X}^{i,j-1} & j > i+2 \end{cases} \qquad (5.31)$$

for $j = 0$,

$$U^{i,j} = \begin{cases} 0 & N = i+1 \\ \widehat{X}^{i,i+1} & N = i+2 \\ \widehat{X}^{i,i+1} + \cdots + \widehat{X}^{i,N-1} & N > i+2 \end{cases} \qquad (5.32)$$

In order to derive the workload distributions, we need firstly to compute the conditional probability $(\widehat{x}_k^{i,j} \mid \overline{X}^i = m \ \wedge \ U^{i,j} = v)$. According to the expression, m transport units are loaded at station i, of which v are already delivered to their destinations before reaching station j. Given this condition, the expression stands for the probability that k units will be delivered to station j. Before we present the equation to compute this probability, we illustrate firstly the calculation principle on a numerical example. Let us assume that an arbitrary transport unit, that is picked up at station 0 is delivered to stations 1, 2, and 3 with probabilities 0.4, 0.25, and 0.35, respectively. If the quantity picked up at station 0 ($\overline{X}^0$) was six transport units and two of them were delivered to station 1 ($\widehat{X}^{0,1} = 2$), then the probability that an arbitrary unit among the remaining four transport units has station 2 as its destination is computed as $P = \frac{0.25}{(1-0.4)}$. Having computed this probability, we can now determine, for instance, the probability that three of four remaining transport units are delivered to station 2:

$$P(\widehat{X}^{0,2} = 3 \mid \overline{X}^0 = 6 \wedge U^{0,2} = 2) = \binom{6-2}{3} \cdot \widetilde{P}^3 \cdot \left(1 - \widetilde{P}\right)^1$$

Subsequently, we attain the general expression:

$$P(\widehat{X}^{i,j} = k \mid \overline{X}^i = m \wedge U^{i,j} = v) = \binom{m-v}{k} \cdot \widetilde{P}^k \cdot (1-\widetilde{P})^{m-v-k} \quad (5.33)$$

where the probability $\widetilde{P}$ is computed as follows:
for $i < j$,

$$
\widetilde{P} = \begin{cases}
p^{i,j} & j = i+1 \\
\frac{p^{i,j}}{1-p^{i,i+1}} & j = i+2 \wedge p^{i,i+1} \neq 1 \\
0 & j = i+2 \wedge p^{i,i+1} = 1 \\
\frac{p^{i,j}}{1-(p^{i,i+1}+\cdots+p^{i,j-1})} & j > i+2 \wedge (p^{i,i+1}+\cdots+p^{i,j-1}) \neq 1 \\
0 & j > i+2 \wedge (p^{i,i+1}+\cdots+p^{i,j-1}) = 1
\end{cases}
$$

$$(5.34)$$

for $j = 0$,

$$
\widetilde{P} = \begin{cases}
p^{i,j} & N = i+1 \\
\frac{p^{i,j}}{1-p^{i,i+1}} & N = i+2 \wedge p^{i,i+1} \neq 1 \\
0 & N = i+2 \wedge p^{i,i+1} = 1 \\
\frac{p^{i,j}}{1-(p^{i,i+1}+\cdots+p^{i,N-1})} & N > i+2 \wedge (p^{i,i+1}+\cdots+p^{i,N-1}) \neq 1 \\
0 & N > i+2 \wedge (p^{i,i+1}+\cdots+p^{i,N-1}) = 1
\end{cases}
$$

$$(5.35)$$

As the next, we consider the independent quantities. A new variable WI^i is introduced, which is the sum of independent quantities listed in the i-th line in figure 5.6. The distribution of this variable is given by

$$
wi^i = \widehat{x}^{i+1,i} \otimes \cdots \otimes \widehat{x}^{N-1,i} \tag{5.36}
$$

Now, the derivation of WL^2 is shown. We rewrite equation 5.29 in terms of (conditional) probabilities.

$$P(WL^2 = \text{a+b+} \cdots + \text{g}) = \sum_{a=0}^{K^0} \overline{x}_a^0 \sum_{b=0}^{a} (\widehat{x}_b^{0,1} \mid \overline{X}^0 = a \wedge U^{0,1} = 0)$$

$$\sum_{c=0}^{WI_{max}^1} wi_c^1 \sum_{d=0}^{K^1} (\overline{x}_d^1 \mid WL^1 = \text{a+b+c}) \sum_{e=0}^{a-b} (\widehat{x}_e^{0,2} \mid \overline{X}^0 = a \wedge U^{0,2} = b)$$

$$\sum_{f=0}^{d} (\widehat{x}_f^{1,2} \mid \overline{X}^1 = d \wedge U^{1,2} = 0) \sum_{g=0}^{WI_{max}^2} wi_g^2 \tag{5.37}$$

The variables WI^1 and WI^2 that appear in equation 5.37 are not necessarily independent of each other. For a milkrun system with e.g. 4 stations, $WI^1 = \widehat{X}^{2,1} + \widehat{X}^{3,1}$ and $WI^2 = \widehat{X}^3$. Obviously, the quantities $\widehat{X}^{3,1}$ and $\widehat{X}^{3,2}$ are correlated, as they originate from the picked quantity at station 3 in an earlier tour. However, we omit dependencies between quantities from an earlier tour. Similarly, we can extend our approach for WL^i:

$$P(WL^i = \text{a+b+} \cdots + \text{l}) = \tag{5.38}$$

$$\sum_{a=0}^{K^0} \overline{x}_a^0 \sum_{b=0}^{a} (\widehat{x}_b^{0,1} \mid \overline{X}^0 = a \wedge U^{0,1} = 0) \sum_{c=0}^{WI_{max}^1} wi_c^1$$

$$\sum_{d=0}^{K^1} (\overline{x}_d^1 \mid WL^1 = a + b + c) \sum_{e=0}^{a-b} (\widehat{x}_e^{0,2} \mid \overline{X}^0 = a \wedge U^{0,2} = b)$$

$$\sum_{f=0}^{d} (\widehat{x}_f^{1,2} \mid \overline{X}^1 = d \wedge U^{1,2} = 0) \sum_{g=0}^{WI_{max}^2} wi_g^2 \cdots \cdots$$

$$\sum_{h=0}^{K^{i-1}} (\overline{x}_h^{i-1} \mid WL^{i-1} = a + \cdots) \sum_{j=0}^{a-b-\cdots} (\widehat{x}_j^{0,i} \mid \overline{X}^0 = a \wedge U^{0,i} = \cdots)$$

$$\sum_{k=0}^{\cdots} (\widehat{x}_k^{i-1,i} \mid \overline{X}^{i-1} = \cdots \wedge U^{i-1,i} = \cdots) \sum_{l=0}^{WI_{max}^i} wi_l^i$$

103

Distributions of the picked quantities

After the distribution of WL^i is computed, we need to compute the conditional distribution of the picked quantity $(\bar{x}^i \mid WT^i = m)$ to proceed to derive WL^{i+1}. Firstly, we compute the queue state conditioned on the workload, which is the sum of X^{+i} and $(G^i \mid WL^i = n)$:

$$(X^i \mid WL^i = n) = X^{+i} + (G^i \mid WL^i = n) \tag{5.39}$$

Note that X^{+i} is the number of units beyond the limit at the start of the previous loading process at station i and independent of the workload realized in the current tour. We get the distribution x^{+i} as in equations 5.13 and 5.14. Subsequently, we derive the distribution $(g^i \mid WL^i = n)$ as follows:

$$P(G^i = m \mid WL^i = n) = \sum_{t=0}^{\infty} (g_m^i \mid C^i = t) \cdot P(C^i = t \mid WL^i = n) \tag{5.40}$$

The expression $P(G^i = m \mid C^i = t)$ in equation 5.40 is obtained following equations 5.16 and 5.17. In order to attain the expression $P(C^i = t \mid WL^i = n)$, we firstly calculate the cycle segment distribution dependent on the workload. If the workload till the loading process at station i was n, the cycle segment is the sum of the service time for n transport units and the driving time till station i. Hence,

$$(cs^i \mid WL^i = n) = b^{n\otimes} \otimes s^0 \otimes \cdots \otimes s^{i-1} \tag{5.41}$$

As illustrated in figure 5.4, cycle time in an arbitrary n^{th} tour is given by the following expression:

$$(C^{i,n} \mid WL^{i,n}) = (CS^{i,n} \mid WL^{i,n}) + T - CS^{i,n-1} \tag{5.42}$$

Note that the cycle segment at station i in the nth tour is dependent on the workload till station i in the nth tour. On the other hand, the cycle segment observed in the previous tour is independent of the workload realized in the actual tour. Considering this, we compute an auxiliary distribution by shifting the distribution $(cs^i \mid WL^i)$ by T units:

$$P(\tilde{CS}^i = m + T \mid WL^i = n) = P(CS^i = m \mid WL^i = n) \tag{5.43}$$

Then, we get the cycle segment distribution:

$$P(CS^i = a) = \sum_{n=0}^{\infty} P(CS^i = a \mid WL^i = n) \cdot P(WL^i = n) \tag{5.44}$$

With the negative convolution operation, we attain the following auxiliary distribution:

$$(\tilde{c}^i \mid WL^i = n) = (\tilde{cs}^i \mid WL^i = n) \otimes -cs^i \tag{5.45}$$

Since we excluded the possibility of an overtake, we bound this auxiliary distribution to zero (see equation 2.13) and attain the cycle time distribution dependent on the workload:

$$(c^i \mid WL^i = n) = \Pi_0[(\tilde{c}^i \mid WL^i = n)] \tag{5.46}$$

Having computed the distributions x^{+i} and $(g^i \mid WL^i = n)$, we can now get the queue state dependent on the workload:

$$(x^i \mid WL^i = n) = x^{+i} \otimes (g^i \mid WL^i = n) \tag{5.47}$$

Finally, the distribution $(\overline{x}^i \mid WL^i = n)$ corresponds to the distribution $(x^i \mid WL^i = n)$ with an upper bound K^i (see equation 2.13 for the pi-operator). Thus, we obtain

$$(\overline{x}^i \mid WL^i = n) = \Pi^{K^i}[(x^i \mid WL^i = n)] \tag{5.48}$$

The distribution $(\overline{x}^i \mid WL^i = n)$ will be used to compute the distribution of the workload till station $(i+1)$ as explained beforehand. Note that $(\overline{x}^0 \mid WL^0 = n) = \overline{x}^0$, as the workload till station 0 is always zero.

Cycle time distributions

In equation 5.46, we have already computed the conditional cycle time distribution. With the law of total probability, we get the cycle time

$$P(C^i = t) = \sum_{n=0}^{\infty} P(C^i = t \mid WL^i = n) \cdot P(WL^i = n) \tag{5.49}$$

Tour time distribution

In order to calculate the tour time distribution (tt), we proceed as follows. Firstly, we compute the total workload in a tour (TWL), which stands for the sum of picked and delivered quantities in a tour. Recall that we consider all the quantities displayed in the first i lines of figure 5.6 to derive the workload WL^i. Analogous to that, TWL is the sum of all the quantities listed in the figure. Consequently, the derivation of TWL is similar to the derivation of the workload WL^i. We consider here all the quantities listed in the figure and the possible dependencies between them for the derivation of TWL. We again use conditional probabilities to account for the dependent variables. Thus, we attain:

$$P(TWL = \text{a+b+}\cdots+\text{p}) = \tag{5.50}$$

$$\sum_{a=0}^{K^0} \overline{x}_a^0 \sum_{b=0}^{a} (\widehat{x}_b^{0,1} \mid \overline{X}^0 = a \wedge U^{0,1} = 0) \sum_{c=0}^{WI_{max}^1} wi_c^1$$

$$\sum_{d=0}^{K^1} (\overline{x}_d^1 \mid WL^1 = a + b + c) \sum_{e=0}^{a-b} (\widehat{x}_e^{0,2} \mid \overline{X}^0 = a \wedge U^{0,2} = b)$$

$$\sum_{f=0}^{d} (\widehat{x}_f^{1,2} \mid \overline{X}^1 = d \wedge U^{1,2} = 0) \sum_{g=0}^{WI_{max}^2} wi_g^2 \ldots\ldots\ldots\ldots$$

$$\sum_{m=0}^{K^{N-1}} (\overline{x}_m^{N-1} \mid WL^{N-1} = a + \cdots + l)\ldots\ldots\ldots$$

$$\sum_{o=0}^{m} (\widehat{x}_o^{N-1,0} \mid \overline{X}^{N-1} = m \wedge U^{N-1,0} = \cdots) \sum_{p=0}^{WI_{max}^N} wi_p^N$$

The variables WI^{N-1} and WI^N are always equal to zero (note that there are no independent quantities in the $(N-1)^{th}$ and N^{th} lines in figure 5.6). So the expressions wi_l^{N-1} and wi_p^N can be omitted.

Once, the distribution of TWL is known, the distribution of the tour time can be computed with following expression.

$$P(TT = m + d) = \sum_{n=0}^{\infty} \sum_{m=0}^{n \cdot b_{max}} P(TWL = n) \cdot b_m^{n\otimes} \sum_{d=0}^{\infty} s_d^* \qquad (5.51)$$

where the distribution of the total driving time is given by $s^* = s^0 \otimes s^1 \otimes \cdots \otimes s^{N-1}$.

Queue state and waiting time distributions

Having improved the cycle time distributions, it is possible to improve the distributions of the queue states and the waiting times. The queue state distributions are improved in accordance with the following equation. In order to improve the queue states, we firstly compute the distribution g^i with the improved cycle time distribution:

$$P(G^i = m) = \sum_{t=0}^{\infty} P(G^i = m \mid C^i = t) \cdot c_t^i \qquad (5.52)$$

The conditional probability $P(G^i = m \mid C^i = t)$ in the previous equation is derived as given in equations 5.16 and 5.17. Eventually, the new queue state distribution can be computed by means of convolution.

$$x^i = x^{+i} \otimes g^i \qquad (5.53)$$

Waiting time distributions can be improved by employing the new cycle time distributions. For this purpose, we follow the method introduced in section 5.1.7.

A note on the improvement algorithm

In equations 5.51 and 5.51, we displayed the distributions of the workloads (WL^i) and the total workload (TWL). The expressions were formulated under the most general assumption that each station delivers goods to every other station in the system and receives goods from all the other stations. In reality, such a case is rare. That's

107

why, expressions are simplified to a great extent for practical cases and the algorithm performs efficiently. Furthermore, the improvement algorithm can be simplified by considering only the dependency between the transferred quantity $\widehat{X}^{i,j}$ on the picked quantity $\overline{X}^i$. Thus, the expression $(\widehat{x}_k^{i,j} \mid \overline{X}^i = s \wedge U^{i,j} = l)$ in the equations can be replaced by

$$(\widehat{x}_k^{i,j} \mid \overline{X}^i = s)$$ as a simplification.

5.1.9. Numerical Results

As we have presented approximate methods for the tour time, cycle time, queue state, and the waiting time distributions, we make here a note on the quality of the basic and improvement algorithms introduced. For a wide range of parameters, the algorithms yield very accurate results for the higher moments and the quantiles of all the measures and generally the computed distributions follow the actual distributions accurately. However, the computed squared coefficients of variation or other variability measures may show significant deviations, especially for the tour time and the cycle times computed by the basic algorithm and possibly some deviations for the waiting times. The deviations are higher especially for the cases, in which an arbitrary transport unit, picked up at a given station i, has a deterministic destination, thus $p^{i,j} = \{0,1\}$ for all transport relations. The reason is that the basic algorithm does not take the correlation between the picked and delivered quantities into account, which becomes significant when $p^{i,j} = \{0,1\}$ for all transport relations. For such systems, the basic algorithm yields some deviations in the tour and cycle time distributions, as expected, whereas the queue state and waiting time distributions follow the actual distributions more accurately. Consequently, it is recommended to improve the tour time and the cycle times, and possibly the queue states and the waiting times with the improvement algorithm introduced in section 5.1.8. However, the improvement offered by the improvement algorithm for the waiting times and the queue states is insignificant compared to the improvement realized for the tour time and the cycle times. Nevertheless, as we will illustrate in this section, the results even from the basic algorithm are sufficiently accurate for the early planning phase.

So as to illustrate the accuracy of our approaches, we study two takted milkrun systems, which we denote as example 1 and example 2. The systems are identical except for the transport matrices. In example 1, probabilities, that make up the transport matrix, is equal to either one or zero, whereas the transport matrix used for example 2 is not restricted to one or zero. In other words, a station in example 2 may deliver to multiple stations. In contrary, a given station in example 1 delivers to exactly one station. The input distributions are displayed in table A.7 and the transport matrices in tables A.8 and A.9 in appendix.

	Example 1				Example 2			
TT	E	scv	$\sigma_{0.95}$	$\sigma_{0.99}$	E	scv	$\sigma_{0.95}$	$\sigma_{0.99}$
sim	88.1418	0.0013	93	95	88.1333	0.0014	94	96
analy (1)	88.1422	0.0007	92	93	88.1422	0.0010	93	95
analy (2)	88.1379	0.0012	93	95	88.1376	0.0015	94	97
$\Delta_{rel.}$ (1)	0.0000	0.4885	0.01	0.02	0.0001	0.2587	0.01	0.01
$\Delta_{rel.}$ (2)	0.0000	0.0727	0.00	0.00	0.0000	0.0460	0.00	0.01
C^1	E	scv	$\sigma_{0.95}$	$\sigma_{0.99}$	E	scv	$\sigma_{0.95}$	$\sigma_{0.99}$
sim	42.0000	0.0008	44	44	42.0000	0.0016	45	46
analy (1)	42.0000	0.0004	43	44	42.0000	0.0013	44	46
analy (2)	42.0000	0.0008	44	44	42.0000	0.0017	45	46
$\Delta_{rel.}$ (1)	0.0000	0.4812	0.02	0.00	0.0000	0.1991	0.02	0.00
$\Delta_{rel.}$ (2)	0.0000	0.0376	0.00	0.00	0.0000	0.0234	0.00	0.00
C^2	E	scv	$\sigma_{0.95}$	$\sigma_{0.99}$	E	scv	$\sigma_{0.95}$	$\sigma_{0.99}$
sim	42.0000	0.0026	45	47	42.0000	0.0052	47	49
analy (1)	42.0000	0.0022	45	46	42.0000	0.0044	47	48
analy (2)	42.0000	0.0026	45	47	42.0000	0.0056	47	49
$\Delta_{rel.}$ (1)	0.0000	0.1732	0.00	0.02	0.0000	0.1442	0.00	0.02
$\Delta_{rel.}$ (2)	0.0000	0.0010	0.00	0.00	0.0000	0.0924	0.00	0.00
C^3	E	scv	$\sigma_{0.95}$	$\sigma_{0.99}$	E	scv	$\sigma_{0.95}$	$\sigma_{0.99}$
sim	42.0000	0.0040	46	48	42.0002	0.0075	48	50
analy (1)	42.0000	0.0034	46	48	42.0004	0.0092	49	51
analy (2)	42.0000	0.0038	46	48	42.0002	0.0070	48	50
$\Delta_{rel.}$ (1)	0.0000	0.1582	0.00	0.00	0.0000	0.2197	0.02	0.02
$\Delta_{rel.}$ (2)	0.0000	0.0650	0.00	0.00	0.0000	0.0663	0.00	0.00

Table 5.1.: Analysis of tour and cycle times in takted milkrun systems; comparison of analytical results with simulation results

For these examples, we compute the performance measures for all stations by means of simulation, the basic and the improvement algorithms. In tables 5.1 to 5.3, we display the mean (E), squared coefficient of variation (scv), 95% and 99%-quantiles of the performance measures are listed. Moreover, relative deviations are displayed. Obvious from the tables that the most deviation occurs for the scv values of the tour time and the cycle times, computed by the basic algorithm. The deviations are much more significant for example 1. Still, the means and the quantiles computed for these measures are highly accurate in both examples. The improvement obtained by the improvement algorithm regarding the scv values is very high. Note that the cycle time distribution at station 0 is not displayed as this is always equal to the takt time. Regarding the waiting time, we see that both algorithms yield some deviations in the mean, scv and quantiles, although the results are still sufficiently accurate. Finally, both algorithms perform well for the queue states.

Besides, we display here the resulting improvement in tour time and cycle time distributions for example 1 in figure 5.7. Waiting time and queue state distributions for example 1 are given in figure 5.8. Note that we just display the waiting time and queue state distributions computed by the basic algorithm, as the ones attained by the improvement algorithm are almost identical, if not more accurate. The resulting distributions for example 2 are displayed in figures A.1 and A.2 in appendix. Obvious from the figures, even the results from the basic algorithm, without the improvement algorithm are sufficiently accurate, whereas the improvement algorithm promises more accurate results.

As the examples differ just in the transport matrices, we compare here the effect of the transport matrix on the performance measures. Tables 5.1, 5.2, and 5.3 show clearly that the effect of the transport matrix on the mean values of the performance measures is minimal. Regarding the variability of the performance measures, the transport matrix has the most effect on the cycle times. The squared coefficients of variation of the cycle times in example 2 are significantly higher than those in example 1, due to the stochastic transport matrix (i.e. an arbitrary transport unit has a stochastic destination). The effect of the transport matrix on the squared coefficients of variation of the queue states, tour time, and waiting times are obviously much less than its effect on the cycle times. Especially scv of the waiting times are very similar to each other in both examples. Eventually, if we compare the quantiles of the performance measures in both cases, we see that the quantiles in example 2 are either equal to or greater than the quantiles in example 1. To sum up, we can state that the variability of the performance measures increases, if the transport matrix is stochastic. But the transport matrix does not have a significant effect on the mean val-

	Example 1				Example 2			
X^0	E	scv	$\sigma_{0.95}$	$\sigma_{0.99}$	E	scv	$\sigma_{0.95}$	$\sigma_{0.99}$
sim	2.2309	0.0466	3	4	2.2289	0.0459	3	4
analy (1)	2.2302	0.0467	3	4	2.2302	0.0467	3	4
analy (2)	2.2302	0.0467	3	4	2.2302	0.0467	3	4
$\Delta_{rel.}$ (1)	0.0003	0.0015	0	0	0.0006	0.0169	0	0
$\Delta_{rel.}$ (2)	0.0003	0.0015	0	0	0.0006	0.0169	0	0
X^1	E	scv	$\sigma_{0.95}$	$\sigma_{0.99}$	E	scv	$\sigma_{0.95}$	$\sigma_{0.99}$
sim	2.8942	0.0754	4	5	2.8993	0.0791	4	5
analy (1)	2.8986	0.0761	4	5	2.9038	0.0808	4	5
analy (2)	2.8986	0.0760	4	5	2.9039	0.0802	4	5
$\Delta_{rel.}$ (1)	0.0015	0.0094	0	0	0.0016	0.0217	0	0
$\Delta_{rel.}$ (2)	0.0015	0.0083	0	0	0.0016	0.0143	0	0
X^2	E	scv	$\sigma_{0.95}$	$\sigma_{0.99}$	E	scv	$\sigma_{0.95}$	$\sigma_{0.99}$
sim	2.6946	0.1148	4	4	2.6946	0.1165	4	4
analy (1)	2.6946	0.1144	4	4	2.6972	0.1166	4	4
analy (2)	2.6946	0.1148	4	4	2.6972	0.1165	4	4
$\Delta_{rel.}$ (1)	0.0000	0.0029	0	0	0.0010	0.0004	0	0
$\Delta_{rel.}$ (2)	0.0000	0.0001	0	0	0.0010	0.0003	0	0
X^3	E	scv	$\sigma_{0.95}$	$\sigma_{0.99}$	E	scv	$\sigma_{0.95}$	$\sigma_{0.99}$
sim	1.9996	0.0277	3	3	1.9999	0.0366	3	3
analy (1)	2.0000	0.0253	3	3	2.0000	0.0345	3	3
analy (2)	2.0000	0.0265	3	3	2.0000	0.0386	3	3
$\Delta_{rel.}$ (1)	0.0002	0.0858	0	0	0.0001	0.0559	0	0
$\Delta_{rel.}$ (2)	0.0002	0.0459	0	0	0.0001	0.0544	0	0

Table 5.2.: Analysis of queue states in takted milkrun systems; comparison of analytical results with simulation results

ues. Subsequently, we study example 2 under different takt times. We compare means and 99%-quantiles of the queue states and the waiting times, which are calculated based on the basic algorithm. The results are visualized in figure 5.9. The mean values of the queue states show an approximately linear trend under different takt times. The quantiles of the queue states show a non-decreasing trend. Only when the takt time is very long so that the station becomes heavily loaded, the 99% quantiles of the queue states increase drastically. This is the case, when the average number of transport units that arrive at the given station in the mean cycle time (which can be approximated by the takt time) approaches the constant limit (see equation 5.1). The same conclusion applies for the means and the 99% quantiles of the waiting times. For instance, for takt times ≥ 42, the change in the means and the 99%-quantiles of the waiting time is drastical for station 0 and station 1, as they become heavily loaded for these takt values. We see that station 3, the means and the quantiles of the waiting times show almost a linear trend. Such a study can be used to dimension the puffers of the visited stations under different takt times or to judge the sojourn time of the transport units and may aid in determining the optimal value of the takt time.

	Example 1				Example 2			
W^0	E	scv	$\sigma_{0.95}$	$\sigma_{0.99}$	E	scv	$\sigma_{0.95}$	$\sigma_{0.99}$
sim	22.5610	0.2766	41	43	22.6087	0.2847	41	43
analy (1)	21.3163	0.3050	40	41	21.3163	0.3050	40	41
analy (2)	21.3163	0.3050	40	41	21.3163	0.3050	40	41
$\Delta_{rel.}$ (1)	0.0552	0.1027	0.02	0.05	0.0572	0.0715	0.02	0.05
$\Delta_{rel.}$ (2)	0.0552	0.1027	0.02	0.05	0.0572	0.0715	0.02	0.05
W^1	E	scv	$\sigma_{0.95}$	$\sigma_{0.99}$	E	scv	$\sigma_{0.95}$	$\sigma_{0.99}$
sim	22.9942	0.2517	41	43	23.0149	0.2536	41	43
analy (1)	21.9401	0.2815	40	42	22.0374	0.2821	40	43
analy (2)	21.9834	0.2813	40	43	22.0699	0.2820	40	43
$\Delta_{rel.}$ (1)	0.0458	0.1187	0.02	0.02	0.0425	0.1121	0.02	0.00
$\Delta_{rel.}$ (2)	0.0440	0.1180	0.02	0.00	0.0411	0.1119	0.02	0.00
W^2	E	scv	$\sigma_{0.95}$	$\sigma_{0.99}$	E	scv	$\sigma_{0.95}$	$\sigma_{0.99}$
sim	22.6495	0.2630	41	44	22.6584	0.2682	41	45
analy (1)	21.5809	0.2889	40	43	21.6692	0.2915	40	44
analy (2)	21.5975	0.2884	40	43	21.7194	0.2829	41	44
$\Delta_{rel.}$ (1)	0.0472	0.0983	0.02	0.02	0.0437	0.0869	0.02	0.02
$\Delta_{rel.}$ (2)	0.0464	0.0966	0.02	0.02	0.0414	0.0546	0.00	0.02
W^3	E	scv	$\sigma_{0.95}$	$\sigma_{0.99}$	E	scv	$\sigma_{0.95}$	$\sigma_{0.99}$
sim	22.0653	0.2872	41	44	22.2362	0.2887	41	45
analy (1)	21.0965	0.3119	40	43	21.1788	0.3158	40	44
analy (2)	21.1055	0.3113	40	43	21.2330	0.3154	41	45
$\Delta_{rel.}$ (1)	0.0439	0.0860	0.02	0.02	0.0476	0.0939	0.02	0.02
$\Delta_{rel.}$ (2)	0.0435	0.0839	0.02	0.02	0.0451	0.0928	0.00	0.00

Table 5.3.: Analysis of waiting times in takted milkrun systems; comparison of analytical results with simulation results

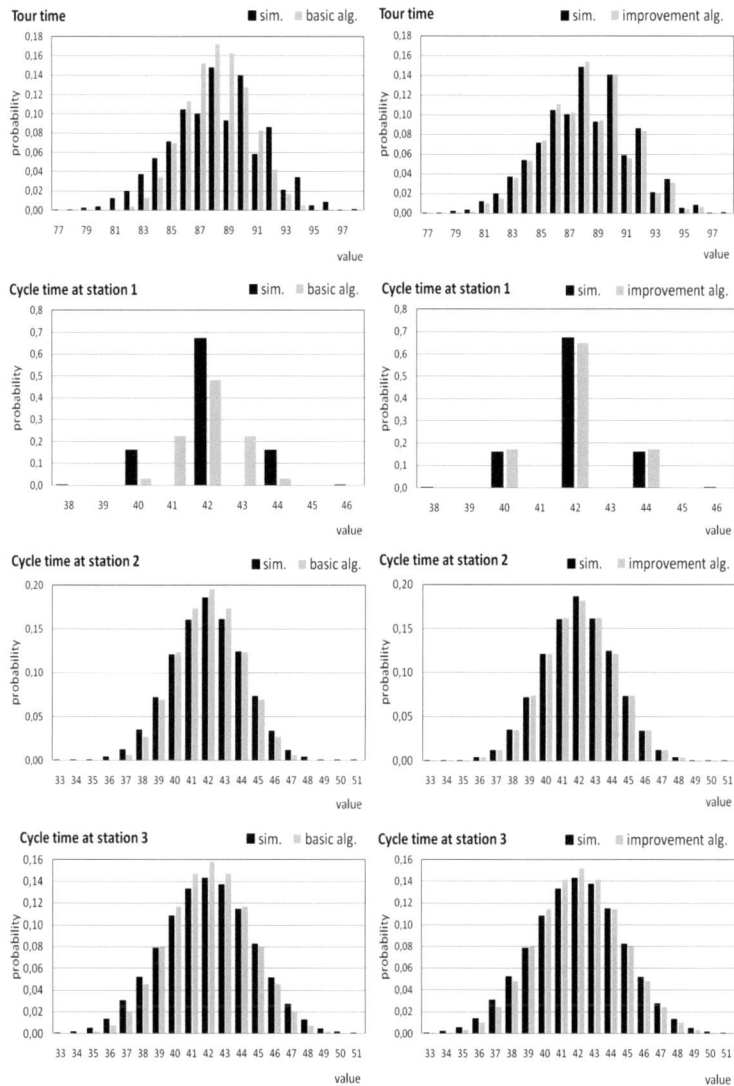

Figure 5.7.: Analysis of takted milkrun systems: tour time and cycle time distributions attained by the basic and the improvement algorithms in comparison to simulation results for example 1

113

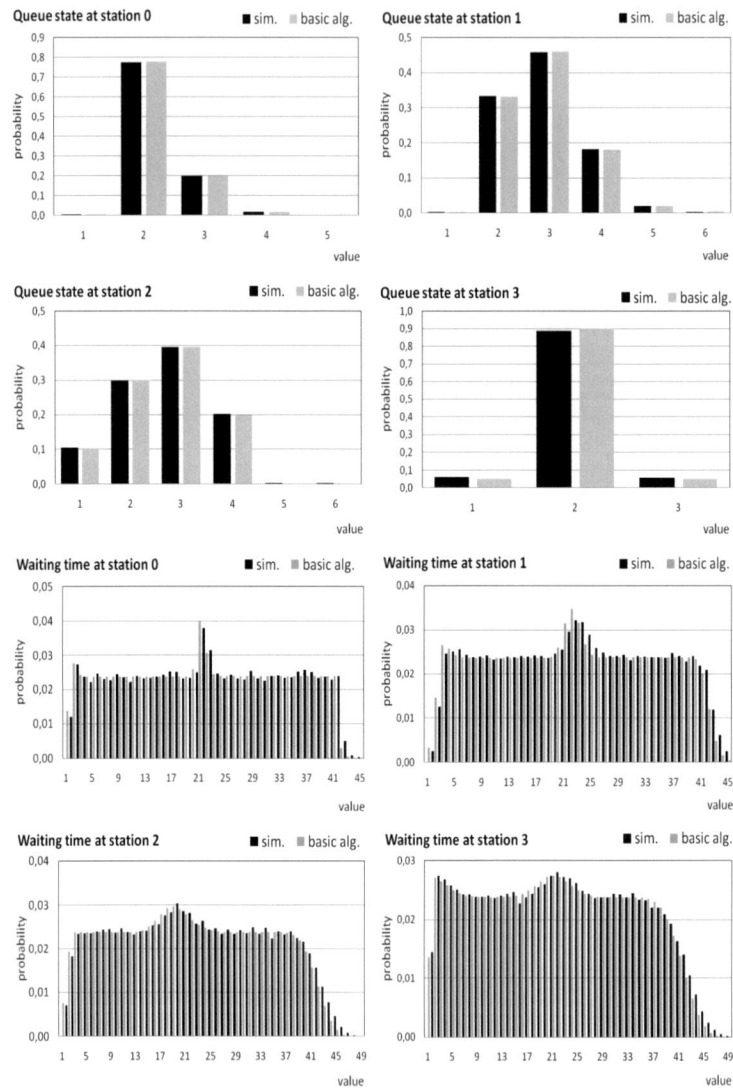

Figure 5.8.: Analysis of takted milkrun systems: queue state and waiting time distributions attained by the basic algorithm in comparison to simulation results for example 1

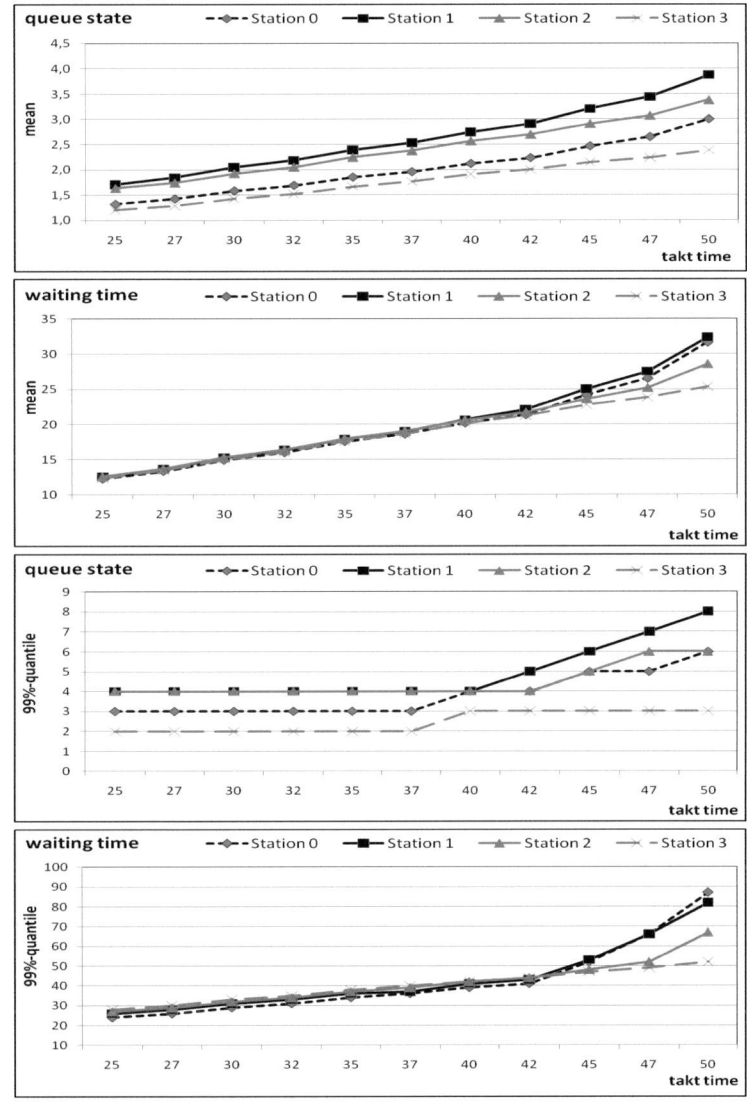

Figure 5.9.: Analysis of takted milkrun systems: mean and 99%-quantiles of the queue states and waiting times for example 2 under different takt times

5.2. Shuttle Milkrun Systems

In this section, we investigate the shuttle milkrun systems with N stations. Different than the takted milkrun system, introduced in the previous section,the vehicles do not depart from the source in fixed time intervals in the shuttle milkrun systems, but there is only one vehicle in the system, which starts a new tour upon the completion of the previous one. Thus, the course of a tour is summarized as follows.

1. load the transport units at station 0 (if station 0 delivers some transport units),
2. drive to the next station,
3. unload the transport units, of which destination is the given station (if there is any),
4. load the transport units at the given station (if the station delivers some transport units),
5. repeat steps 2-4 till station (N-1),
6. drive back to station 0,
7. unload the transport units, of which destination is station 0 (if there is any),
8. go back to step 1.

We allow here also a bidirectional material flow and assume that each station has a loading and an unloading station. The loading process takes place in accordance with the gated limited policy. Thus, the number of transport units, that can be loaded at an arbitrary station, is limited to an arbitrary constant. Furthermore, the transport units, that arrive at the station after the start of the loading process, are not loaded in the current tour. The transport matrix visualizes the values for $p^{i,j}$, which is the probability, that an arbitrary transport unit loaded at station i is delivered to station j.

5.2.1. Queuing System

In this model, the arrival process at an arbitrary station i is characterized by the inter-arrival time A^i and the batch size Y^i. We assume that the variable B is the service time needed either to load or to unload a transport unit. Complying with the notation introduced in the preceding section, the driving time between two successive stations is also denoted by S^i. We assume these variables are discrete and iid variables. For this system, we derive the queue states X^i, waiting time of an arbitrary unit W^i, and the tour time TT. The important variables and parameters are summarized below:

K^i constant limit for the quantity picked up (loaded) at station i in a tour,

S^i driving time between station i and the next station,

A^i inter-arrival time at station i,

Y^i incoming batch size at station i,

R^i residual inter-arrival time at station i,

B loading/unloading time for a transport unit,

X^i queue state at station i, number of transport units, that the vehicle sees immediately before the start of a loading process at station i,

$\overline{X}^i$ number of transport units loaded at station i,

X^{+i} number of transport units beyond the limit at the start of a loading process at station i,

$p^{i,j}$ probability, that an arbitrary transport unit loaded at station i is delivered to station j,

$\widehat{X}^{i,j}$ number of transport units delivered (transferred) from station i to station j,

C^i cycle time at station i, time interval between two successive loading processes at station i,

G^i number of units arrived within a cycle at station i,

CWL^i workload contributed by station i

TWL total workload in a tour, sum of picked and delivered quantities in a tour,

CB total loading and unloading time in a tour,

TT tour time, duration of a tour,

W^i waiting time of a transport unit at station i.

The overall utilization is given by $\rho = \sum_{i=0}^{N-1}(2 \cdot E[Y^i] \cdot E[B])/E[A^i]$. It is noteworthy to mention that the mean service time is multiplied with two, as the service of a single transport unit consists of loading and unloading of the unit.

System is stable only if $\rho < 1$ and the mean number of incoming transport units at station i is less than the limit K^i:

$$\frac{(\sum_{i=0}^{N-1} E(S^i)) \cdot E(Y^i)}{E(A^i)(1 - \rho)} < K^i \qquad \forall i = 0, 1, \cdots, N - 1 \tag{5.54}$$

Note that our analysis applies only for stable systems.

5.2.2. Iterative Algorithm

Before we explain the algorithm for the analysis of milkrun systems, we have to investigate the relationship between the cycle times at different stations. Due to the symmetry of the system, the cycle times at different stations are given by the same expression:

$$C = TWL \cdot B + S^* \tag{5.55}$$

where the total driving time $S^* = S^0 + S^1 + \cdots + S^{N-1}$ and the total workload (TWL) is the sum of all quantities picked or delivered in a cycle, thus:

$$
\begin{aligned}
TWL \quad &= \overline{X}^0 + \widehat{X}^0 + \cdots + \overline{X}^{N-1} + \widehat{X}^{N-1} \\
&= \overline{X}^0 + \widehat{X}^{1,0} + \cdots + \widehat{X}^{N-1,0} \\
&+ \cdots \\
&+ \overline{X}^{N-1} + \widehat{X}^{0,N-1} + \cdots + \widehat{X}^{N-2,N-1}
\end{aligned} \tag{5.56}
$$

Moreover, the tour time is given by the same expression as the cycle time. Hence,

$$T = TWL \cdot B + S^* \tag{5.57}$$

After this brief discussion of cycle times at different stations and the tour time, we ask the following questions:

- Are the distributions of the cycle times at different stations same?

- Does the tour time have the same distribution regardless of how a tour is defined, i.e. which station observes it?

Very interestingly, the answers to both questions are "No". In shuttle systems, the cycle time distributions at arbitrary stations do not have to be exactly the same. Some deviations arise as explained subsequently.

Let us assume a shuttle system with N stations, in which all the transport units loaded at station 2 are unloaded at station 1. Figure 5.10 illustrates the contents of the total workloads in arbitrary cycles for two cases. In the first case, the cycle time is observed at station 0. In this case, the stations are visited in a cycle in the following order: $0 \rightarrow 1 \rightarrow \cdots \rightarrow (N-1) \rightarrow 0 \rightarrow 1 \rightarrow \cdots \rightarrow (N-1) \cdots$. A new cycle starts with the loading process at station 0. In the second case, the cycle time is observed at station 1, thus the sequence, in which the stations are visited, is $1 \rightarrow \cdots \rightarrow (N-1) \rightarrow 0 \rightarrow 1 \rightarrow \cdots \rightarrow (N-1) \rightarrow 0 \cdots$. A new cycle starts when the vehicle starts loading at station 1.

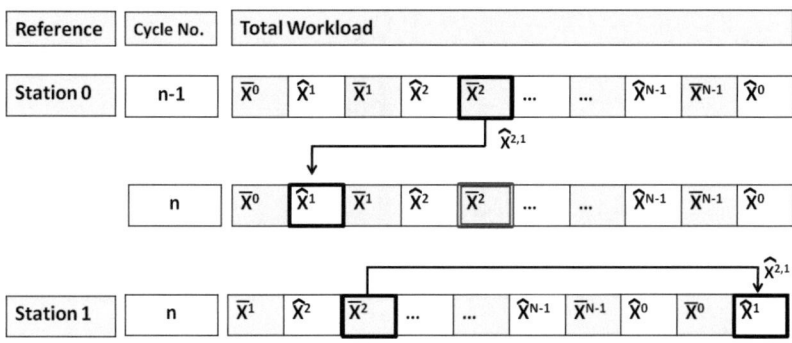

Figure 5.10.: Illustration of the total workloads when the cycle time is observed at different stations

Depending on the arbitrary station, for which the cycle time is computed, the relationship between the loaded quantity at station 2 ($\overline{X}^2$) and the transferred quantity from station 2 to station 1 ($\widehat{X}^{2,1}$) in the current cycle changes. If we observe the cycle time from station 0, these are independent. The reason is that the vehicle visits station 2 after station 1 if the cycle starts at station 0. Thus, $\widehat{X}^{2,1}$ originates from the loaded quantity at station 2 in the previous cycle and is independent of $\overline{X}^2$ realized in the actual cycle. Taking station 0 as the reference point, $\widehat{X}^{2,1}$ can be different than $\overline{X}^2$ in an arbitrary cycle. On the other hand, if we observe the cycle time from station 1, we see that the quantity loaded at station 2 will be unloaded at station 1 in the same cycle, thus, $\widehat{X}^{2,1}$ is always equal to $\overline{X}^2$ in an arbitrary cycle. As the dependencies between the loaded and unloaded quantities may change depending on the reference point, the distributions of the total workload, thus, the cycle times for different stations may look slightly different than each other. Figure 5.11 displays the cycle time distributions for a system with three stations, which are obtained by simulation. Deviations are evident from the figure.

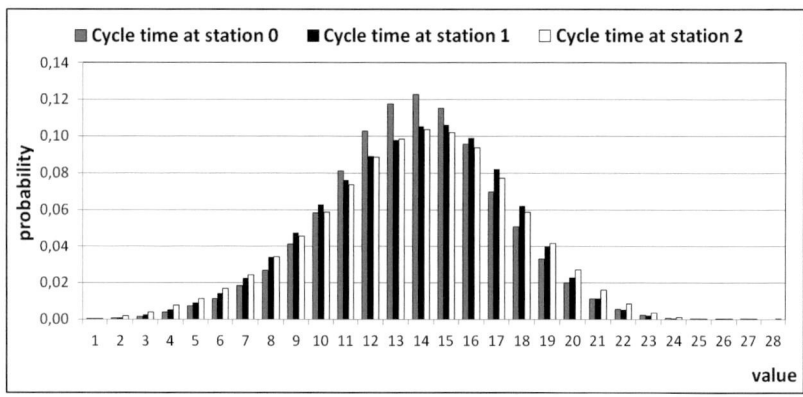

Figure 5.11.: Cycle time distributions for a shuttle milkrun system with three stations

A similar discussion applies for the tour time. The tour time distribution depends also on the reference point. In fact, if we assume e.g. that the tour starts at an arbitrary station i, then the tour time corresponds to the cycle time at station i.

In our analysis, we do not calculate the cycle time distributions at arbitrary stations separately. Instead, we approximate the cycle time distributions at different stations with the cycle time distribution observed at station 0. We denote this distribution as the cycle time (C). Assuming that the tour starts at station 0, the tour time (TT) corresponds also to the cycle time (C).

Analogous to the case analyzed in Dittmann and Hübner (1993), cycle time (C) determines on one hand the queue states (X^i) and on the other hand queue states determine the cycle time. That's why, it is difficult to obtain C and X^i in closed-form expressions. So, we employ an iterative algorithm, of which steps are summarized below:

> 1) Initialize the queue states (X^i) i.e. by setting the system size to zero, and the cycle time C deterministically.
>
> 2) Calculate (X^i) from (C) obtained in the previous iteration (see section 5.2.3) and calculate (C) from (X^i) (see section 5.2.4),
>
> 3) Repeat step 2) until a convergence criterion is fulfilled.

We use in our analysis the absolute difference between the mean cycle time from the actual iteration step and that from the previous iteration step (i.e. $|E(C^{n+1}) - E(C^n)| \leq 0.0001$) as the convergence criterion.

Initialization

The zero-fold convolution of the input variables are assumed to be Dirac distributed with a constant value of zero as in section 5.1.2. Thus, $a_0^{i,0\otimes} = y_0^{i,0\otimes} = b_0^{0\otimes} = 1$ for all stations.

The residual inter-arrival time at an arbitrary station is approximated as given by equation 5.2. Finally, we set the initial cycle time to a deterministic value (e.g. $C = \sum_{i=0}^{N-1} E[S^i]$) and the initial queue states to zero; $x_0^i = 1$.

5.2.3. Queue State Distributions

In this section, we update the distribution of X^i based on the distributions of X^i and C from the previous iteration step. As discussed in section 5.1.4, the queue state is the sum of the transport units beyond the limit at the beginning of the previous loading process (X^{+i}) and the units arrived at the station in the cycle time (G^i). Thus, $X^i = X^{+i} + G^i$.

We compute firstly the number of units beyond the limit ($X^{+i} = \max(0, X^i - K^i)$). The derivation of X^{+i} is the same as explained in section 5.1.4. Thus, we follow equations 5.13 and 5.14.

Subsequently, we derive the number of incoming transport units at station i given the cycle time:

for $k \neq 0$,

$$P(G^i = k \mid C = m) = \sum_{s=1}^{a^i_{max}} r^i_s \sum_{l=1}^{\lceil \frac{m}{a^i_{min}} \rceil} \sum_{g=0}^{m-s} a^{i,(l-1)\otimes}_{m-s-g} \cdot \overline{a^i_{g+1}} \cdot y^{i,l\otimes}_k \quad (5.58)$$

With the sum of steady state probabilities, it yields:

$$P(G^i = 0 \mid C = m) = 1 - \sum_{k=1}^{\infty} P(G^i = k \mid C = m) \quad (5.59)$$

We convolute the conditional distribution ($g^i \mid C = m$) with x^{+i} in order to calculate the conditional distribution ($x^i \mid C = m$):

$$(x^i \mid C = m) = x^{+i} \otimes (g^i \mid C = m) \quad (5.60)$$

Eventually, we determine the queue state at station i by means of the law of total probability:

$$P(X^i = k) = \sum_{m=1}^{\infty} P(X^i = k \mid C = m) \cdot c_m \quad (5.61)$$

Note that we compute the distributions of the queue states for all stations, before we can proceed to calculate the new cycle time distribution.

5.2.4. Cycle and Tour Time Distributions

Having computed the queue states for all stations, we can now determine the distribution of the cycle length, which also corresponds to the tour time. Recall that the queue states and the cycle time are correlated;

e.g. as the queue lengths increase, the cycle time increases. Due to the increased cycle time, queue lengths increase in turn. Thus, there is a correlation between the cycle times realized in successive cycles. Moreover, stations see similar realizations of the cycle time. To account for these, we use the conditional probabilities $(x^i \mid C = m)$ to compute the new cycle time distribution.

We firstly derive the conditional distribution $(\overline{x}^i \mid C = m)$ for every station i ($i : 0, 1, \cdots, N - 1$). At an arbitrary station i, at most K^i transport units are loaded. Hence, we employ the operator Π^M introduced in equation 2.13:

$$(\overline{x}^i \mid C = m) = \Pi^{K^i}[(x^i \mid C = m)] \tag{5.62}$$

Subsequently, we introduce the variable CWL^i, which is the workload contributed by station i in a cycle. That is the sum of transport units loaded at station i and transport units that originate from station i and unloaded at other stations. Thus;

$$CWL^i = \overline{X}^i + \widehat{X}^{i,1} + \widehat{X}^{i,2} + \cdots + \widehat{X}^{i,N-1} + \widehat{X}^{i,0}$$

Similar to the case in section 5.1.8, the delivered quantity $\widehat{X}^{i,j}$ can be dependent on the picked quantity $\overline{X}^i$. If station i is visited before the station j in a cycle, the transferred quantity $\widehat{X}^{i,j}$ depends on the loaded quantity $\overline{X}^i$. For instance, if the loaded quantity at station i was four units, the transferred quantity from this station cannot exceed four units. However, if station i is visited after station j in a cycle, $\widehat{X}^{i,j}$ originates from $\overline{X}^i$ from the previous cycle and is independent of the quantity to be loaded at station i in the current cycle. Thus, $\widehat{X}^{i,j}$ depends on $\overline{X}^i$ given that $(j > i)$ or $(j = 0)$. In order to account for the possible dependency between the loaded quantity at an arbitrary station and the transferred quantities from this station in a cycle, CWL^i comprises two components. The first component $(CWL^{i,I})$ is the sum of quantity picked up at station i and dependent proportion of quantity delivered from station i in a tour. It is given by $CWL^{i,I} = \overline{X}^i + \widehat{X}^{i,i+1} + \cdots + \widehat{X}^{i,N-1} + \widehat{X}^{i,0}$. Thus, it yields

$$P(CWL^{i,I} = n+v \mid C = m) = \sum_{n=0}^{K^i} (\overline{x}_n^i \mid C = m) \sum_{v=0}^{n} \binom{n}{v} p^v (1-p)^{n-v}$$

(5.63)

where p is the probability that a transport unit loaded at station i in this cycle will be unloaded before the cycle ends. Therefore,

$$p = \sum_{\text{all } j} p^{i,j} \qquad \forall j : (j > i) \vee (j = 0)$$

The second component is given by $CWL^{i,II}$ and represents independent proportion of the quantity delivered from station i. This quantity originates from the quantity picked up at station i in the previous cycle. Hence, $CWL^{i,II} = \widehat{X}^{i,1} + \cdots + \widehat{X}^{i,i-1}$. Correspondingly, we compute the following conditional probability:

$$P(CWL^{i,II} = z \mid C = m) = \sum_{n=0}^{K^i} (\overline{x}_n^i \mid C = m) \sum_{z=0}^{n} \binom{n}{z} (1-p)^z p^{n-z}$$

(5.64)

We convolute these conditional distributions and attain the conditional workload for an arbitrary station i.

$$(cwl^i \mid C = m) = (cwl^{i,I} \mid C = m) \otimes (cwl^{i,II} \mid C = m)$$

(5.65)

Subsequently, we convolute the conditional workloads of all stations to attain the conditional total workload in a cycle:

$$(twl \mid C = m) = (cwl^0 \mid C = m) \otimes \cdots \otimes (cwl^{N-1} \mid C = m)$$

(5.66)

The conditional probability $P(CB = k \mid C = m)$ is described as the total loading and unloading time in a cycle, given that the length of the previous cycle equals m time units, and computed as follows:

$$P(CB = k \mid C = m) = \sum_{n=0}^{\infty} P(TWL = n \mid C = m) \cdot b_k^{n\otimes}$$

(5.67)

As the next step, we attain the cycle time distribution conditioned on the cycle time realized in the previous cycle $(c \mid C = m)$:

$$(c \mid C = m) = (cb \mid C = m) \otimes s^0 \otimes \cdots \otimes s^{N-1} \tag{5.68}$$

By applying the law of total probability, we compute the new distribution of the cycle length. The computed cycle time corresponds also to the the tour time.

$$P(C = k) = \sum_{m=1}^{\infty} P(C = k \mid C = m) \cdot c_m \tag{5.69}$$

5.2.5. Waiting Time Distributions

In order to derive the waiting time distribution of an arbitrary transport unit at station i, we proceed the same way as discussed in section 5.1.7. However, cycle times at different stations in takted milkrun systems have different distributions given by C^i. In contrast, the same cycle time distribution (C) applies for all stations in shuttle milkrun systems. That's why, the distributions c^i found in the given equations are replaced by the distribution (c) for shuttle milkrun systems.

5.2.6. Numerical Results

We demonstrate here the accuracy of our approach with two examples; example 1 and example 2 (see tables A.10 to A.12 for input data). In example 1, we consider a system with three stations, in which each station serves exactly one station. In example 2, we consider a larger system with five stations, and each station may deliver to multiple stations. The mean (E), squared coefficient of variation (scv), 95%, 97.5%, and 99%-quantiles for the tour time, queue state, and the waiting time are compared to simulative results in tables 5.4 and 5.5. The relative deviations are also given.

Example 1					
TT	E	scv	$\sigma_{0.95}$	$\sigma_{0.975}$	$\sigma_{0.99}$
sim	38.6394	0.0074	44	45	46
analy	38.6367	0.0060	43	44	45
$\Delta_{rel.}$	0.0001	0.1905	0.0227	0.0222	0.0217
X^0	E	scv	$\sigma_{0.95}$	$\sigma_{0.975}$	$\sigma_{0.99}$
sim	4.3439	0.1821	8	9	11
analy	4.1960	0.1624	7	8	10
$\Delta_{rel.}$	0.0340	0.1085	0.1250	0.1111	0.0909
X^1	E	scv	$\sigma_{0.95}$	$\sigma_{0.975}$	$\sigma_{0.99}$
sim	3.0426	0.0238	4	4	4
analy	3.0422	0.0241	4	4	4
$\Delta_{rel.}$	0.0001	0.0127	0.0000	0.0000	0.0000
X^2	E	scv	$\sigma_{0.95}$	$\sigma_{0.975}$	$\sigma_{0.99}$
sim	3.2191	0.2044	6	6	6
analy	3.3172	0.2299	6	6	6
$\Delta_{rel.}$	0.0305	0.1249	0.0000	0.0000	0.0000
W^0	E	scv	$\sigma_{0.95}$	$\sigma_{0.975}$	$\sigma_{0.99}$
sim	30.9303	0.3900	67	77	97
analy	26.5555	0.3509	55	65	77
$\Delta_{rel.}$	0.1414	0.1001	0.1791	0.1558	0.2062
W^1	E	scv	$\sigma_{0.95}$	$\sigma_{0.975}$	$\sigma_{0.99}$
sim	21.1760	0.2452	38	40	41
analy	20.0978	0.2713	37	39	40
$\Delta_{rel.}$	0.0509	0.1062	0.0263	0.0250	0.0244
W^2	E	scv	$\sigma_{0.95}$	$\sigma_{0.975}$	$\sigma_{0.99}$
sim	24.8409	0.2867	48	53	59
analy	24.3492	0.3104	47	53	64
$\Delta_{rel.}$	0.0198	0.0827	0.0208	0.0000	0.0847

Table 5.4.: Analysis of shuttle milkrun systems; comparison of analytical results with simulation results for example 1

The tables show that the results for example 2 are more accurate than the results for example 1. As the case with the takted systems, the transport matrix is the most important factor regarding the quality of results. The algorithm performs the best when a transport unit loaded at station i has a stochastic destination. Besides, we display in figure 5.12 the distributions for example 1 (for example 2 see figures A.3, A.4, and A.5 in appendix). Obviously, the accuracy of the method is sufficient for the early planning phase for both cases.

Example 2					
TT	E	scv	$\sigma_{0.95}$	$\sigma_{0.975}$	$\sigma_{0.99}$
sim	187.9772	0.0016	200	202	205
analy	187.9771	0.0015	200	202	205
$\Delta_{rel.}$	0.0000	0.0561	0.0000	0.0000	0.0000
X^0	E	scv	$\sigma_{0.95}$	$\sigma_{0.975}$	$\sigma_{0.99}$
sim	13.7640	0.0208	17	18	19
analy	13.7127	0.0197	17	18	19
$\Delta_{rel.}$	0.0037	0.0525	0.0000	0.0000	0.0000
X^1	E	scv	$\sigma_{0.95}$	$\sigma_{0.975}$	$\sigma_{0.99}$
sim	12.3924	0.0083	14	15	15
analy	12.3817	0.0081	14	15	15
$\Delta_{rel.}$	0.0009	0.0261	0.0000	0.0000	0.0000
X^2	E	scv	$\sigma_{0.95}$	$\sigma_{0.975}$	$\sigma_{0.99}$
sim	6.3723	0.0062	7	7	7
analy	6.3721	0.0064	7	7	7
$\Delta_{rel.}$	0.0000	0.0316	0.0000	0.0000	0.0000
X^3	E	scv	$\sigma_{0.95}$	$\sigma_{0.975}$	$\sigma_{0.99}$
sim	15.0392	0.0047	16	16	16
analy	15.0627	0.0050	16	16	18
$\Delta_{rel.}$	0.0016	0.0612	0.0000	0.0000	0.1250
X^4	E	scv	$\sigma_{0.95}$	$\sigma_{0.975}$	$\sigma_{0.99}$
sim	10.7456	0.0241	14	14	15
analy	10.7628	0.0243	14	14	15
$\Delta_{rel.}$	0.0016	0.0091	0.0000	0.0000	0.0000
W^0	E	scv	$\sigma_{0.95}$	$\sigma_{0.975}$	$\sigma_{0.99}$
sim	105.8288	0.2455	185	194	208
analy	103.6650	0.2487	183	191	201
$\Delta_{rel.}$	0.0204	0.0132	0.0108	0.0155	0.0337
W^1	E	scv	$\sigma_{0.95}$	$\sigma_{0.975}$	$\sigma_{0.99}$
sim	101.0638	0.2563	180	185	193
analy	99.9286	0.2614	179	185	191
$\Delta_{rel.}$	0.0112	0.0200	0.0056	0.0000	0.0104
W^2	E	scv	$\sigma_{0.95}$	$\sigma_{0.975}$	$\sigma_{0.99}$
sim	97.5112	0.2913	179	183	190
analy	96.3401	0.2993	178	183	190
$\Delta_{rel.}$	0.0120	0.0273	0.0056	0.0000	0.0000
W^3	E	scv	$\sigma_{0.95}$	$\sigma_{0.975}$	$\sigma_{0.99}$
sim	101.4795	0.2437	179	184	190
analy	101.1621	0.2470	180	184	191
$\Delta_{rel.}$	0.0031	0.0136	0.0056	0.0000	0.0053
W^4	E	scv	$\sigma_{0.95}$	$\sigma_{0.975}$	$\sigma_{0.99}$
sim	103.4941	0.2596	184	191	204
analy	102.4481	0.2634	184	191	202
$\Delta_{rel.}$	0.0101	0.0146	0.0000	0.0000	0.0098

Table 5.5.: Analysis of shuttle milkrun systems; comparison of analytical results with simulation results for example 2

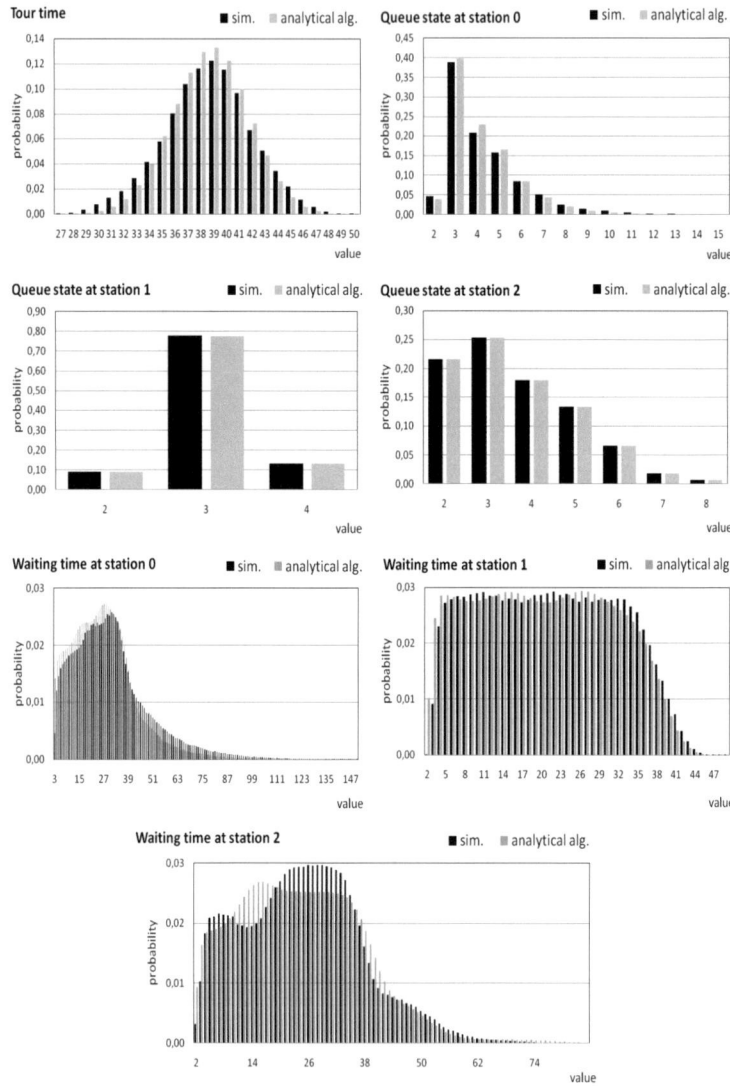

Figure 5.12.: Analysis of shuttle milkrun systems: tour time, queue state, and waiting time distributions attained by the analytical algorithm in comparison to simulation results for example 1

6. Terminal Consolidation

6.1. Introduction

The implementation of conventional direct transport processes entails a direct and unobstructed flow of material between the source and the drain. However, in a transport network, in which all the nodes are connected to each other directly, the number of routes cannot be handled economically. Regarding this problem, a solution to be considered is the implementation of terminal consolidation.

Terminal consolidation is a spatial form of consolidation, which involves construction of transit terminals, over which the material streams flow. Thus, the nodes are not directly connected to each other, but rather to one or more central nodes (terminals). In this way, the number of connections is reduced drastically. For a network of n nodes, direct transport results in $\frac{n \cdot (n-1)}{2}$ routes and the complexity is $O(n^2)$. Assuming that each node is connected to exactly one terminal, this number reduces to $(n-1)$ and the complexity is given by $O(n)$. If, for instance, there are five nodes in the network, the number of connections decreases from ten to four with the implementation of terminal consolidation.

A typical example of terminal consolidation is the hub-and-spoke network. In the terminology of the hub-and-spoke networks, hubs describe the terminals and spokes the routes from/to terminals. Figure 6.1 displays a transport network before and after the implementation of the hub-and-spoke form. Reduction in number of connections is evident. Owing to the reduced number of routes with more frequent service, transport capacities are utilized much more efficiently in this form. However, a drawback of this form is the longer transport ways, as the material must be routed through a number of hubs on the way to its destination. Furthermore, the transport processes become less reliable since e.g. delays in some routes or capacity problem in a hub may yield unexpected consequences for the whole network. Hence, precise analysis is needed to design and operate hub-and-spoke networks.

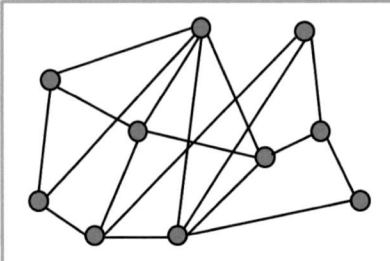 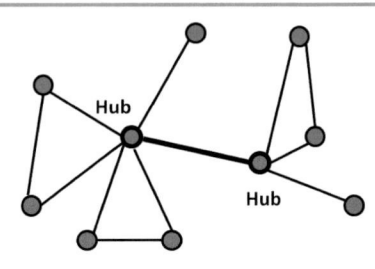

Figure 6.1.: Restructuring of a transport network in hub-and-spoke configuration: left: original network; right: restructured network

For the analysis of hub-and-spoke networks, we propose here the discrete time network analysis. For this purpose, we analyze a hub-and-spoke network by means of the models introduced in this work.

In a network, the departure processes of the preceding nodes make up the arrival process at a given node. Discrete time analysis can be implemented in open networks, easily by computing the inter-departure distribution at a node and employing this distribution as the inter-arrival time at the next node. While doing so, it is assumed that the nodes are stochastically independent. However, it is well known that the departure process is generally correlated (see Schleyer (2007, pp. 116)). This correlation is not considered, if the next model element assumes iid inter-arrival times, which is a very common assumption in most queuing models. Thus, the results may show some deviations.

We present in section 6.2 a numerical study of the material flow in a hub-and-spoke transport network. The objective there is to evaluate the suitability of the discrete time analysis for the performance measure of the hub-and-spoke networks.

6.2. Numerical Case: Analysis of a Hub-and-Spoke Network

The investigated network is illustrated in figure 6.2. In this network, there are two terminals and seven stations. Basically, shipments from

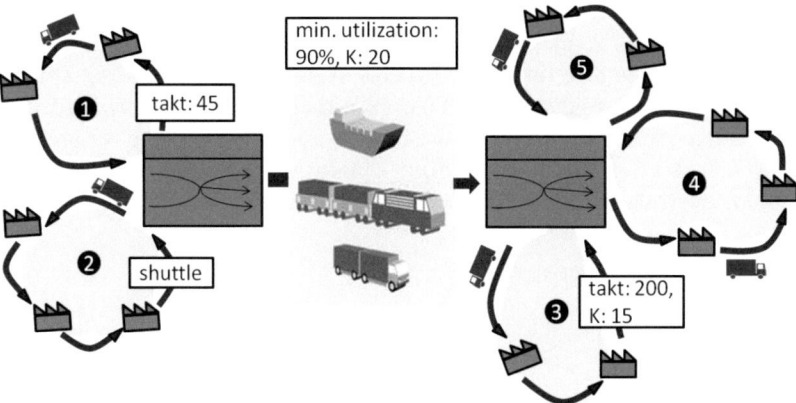

Figure 6.2.: Illustration of the material flow in the studied hub-and-spoke network

two regions are collected and delivered to terminal 1. Region 1 consists of station 1 and station 2 and is served in takted intervals of 45 time units. Region 2 comprises three stations, that are served in shuttle milkrun tours. At terminal 1, shipments from these regions are stored until the required minimum utilization of 90% is satisfied, where the capacity of the employed vehicle is limited to 20 units. Thereafter, the collected transport units are transferred to terminal 2. At terminal 2, the shipments are sorted for different regions. In our analysis, we investigate Region 3, to which station 6 and station 7 belong. Vehicles are dispatched to this region as follows. A maximum collecting time ($tout = 200$) is defined. The vehicle departs either the maximum collecting time elapses or when the vehicle is fully loaded. Here, the vehicle has a capacity of 15 transport units.

For the sake of simplicity, we assume that loading or unloading of a single transport unit takes one unit time in milkrun systems. Furthermore, we made also some assumptions, regarding the handling processes at terminals. At a terminal, the incoming vehicle is completely unloaded as the first. Only when it is completely empty, the unloaded quantity can be loaded on the waiting vehicle. Additionally, transport units are unloaded singly at terminal 1, each takes one time unit. On the contrary, the loading process at terminal 1 takes ten time units per vehicle, regardless of the shipment size. Similarly, loading or unloading of a vehicle at terminal 2 takes five time units. The assumptions of lump-sum loading or unloading times for vehicles at terminals provide significant simplification, in contrast to the assumption of loading/unloading times per transport unit. The transport matrix and other input data are displayed in section A.4in appendix. Finally, we assume that a vehicle cannot overtake another vehicle, that started the route earlier.

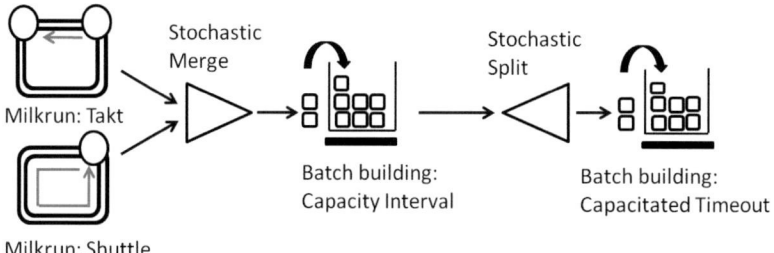

Figure 6.3.: Queuing network for the studied hub-and-spoke network

Using some of the model elements presented earlier, we build the queuing network for the material flow described above. The queuing network is illustrated in figure 6.3. We use the takted and shuttle milkrun models to analyze region 1 and region 2, respectively. The arrival process at terminal 1 is modeled with the stochastic merge models introduced by (Furmans 2004a) and (Schleyer 2007, pp. 112). Subsequently, the transport process from terminal 1 to terminal 2 is represented by the capacity interval batch building model. Sorting at terminal 2 is computed with a split element, that is explained later in this section. Finally, the capacitated timeout batch building model is employed for the transport process from terminal 2 to region 3.

The material flow in the investigated network is obviously not a simple unidirectional one. In order to explain this in further detail, let us investigate the material flow at terminal 1. The transport units collected in region 1 and region 2 are delivered to terminal 1. But there is also a material flow from terminal 1 to the stations in these regions. For instance, empty containers can be collected at terminal 1 and sent back to the stations in these regions. Besides this flow, there is also a material flow from terminal 1 to terminal 2. For the study of the material flow in the network, some additional computation is needed to regulate the interface between the model elements. This is the case, when the model element does not compute the measure needed directly. We explain here the computational steps briefly.

Analysis of the arrival process at terminal 1: here we consider two arrival streams. Arrival stream 1 is the material flow from region 1 to terminal 1. The inter-arrival time of vehicles for the takted milkrun systems is not equal to the tour time (TT). We calculate it additionally. Let us assume that the nth vehicle needed TT^n time units to return back to terminal 1. The $(n+1)^{th}$ vehicle starts the tour a takt time (T) later than the previous one and returns in TT^{n+1} time units to terminal 1. Then, the inter-arrival time is given by $A^{n+1} = \max\{TT^{n+1} + T - TT^n, 0\}$ (note that we exclude the possibility of a overtake between the vehicles). Neglecting the dependency between the courses of two successive tours, the inter-arrival time is calculated as follows:

$$a = \Pi_0[\Delta_T[tt] \otimes -tt] \tag{6.1}$$

Operators Δ_m and Π_m are explained in section 2.2. The incoming shipment size from this stream is the unloaded quantity at terminal 1. That is computed readily by the takted milkrun model element. Thus, no additional computation is needed. For arrival stream 2 from region 2 to terminal 1, we employ the tour time as the inter-arrival time and the unloaded quantity at terminal 1 as the incoming batch size.

After this, we need to compute the inter-arrival time and batch size for the merged stream (for the computation of the inter-arrival time see (Furmans 2004a) and for the batch size see Schleyer (2007, p. 112)). These distributions are used as inputs for the capacity interval model.

Analysis of the arrival process at terminal 2: as the driving time between the terminals (see table A.17) and the handling time at terminal 2 are constant, the arrival process at terminal 2 is described by the inter-departure and the departing batch size distributions from the capacity interval model element.

Analysis of sorting and transport processes to region 3: firstly, we compute the quantity sorted from terminal 2 to region 3 by the split element given by:

$$P(Y^s = m) = \sum_{n=m}^{\infty} P(Y = n) \binom{n}{m} p^m (1-p)^{n-m} \tag{6.2}$$

where Y is the incoming batch size at terminal 2 and Y^s is the quantity sorted to region 3. The probability p is the probability that a transport unit is transferred from terminal 2 to region 3, which is presented in the transport matrix (see table A.18). Afterwards, we employ the capacitated timeout model. Input distributions are the sorted quantity from the previous equation and the inter-arrival time at terminal 2.

As the performance measure, we analyze the sojourn time of an arbitrary transport unit. Specifically, sojourn times needed from station 2 to station 7 and from station 4 to station 7 are computed. We define the sojourn time as the time interval from the arrival of a transport unit at the source station till its arrival at the drain station. As mentioned above, not every component of the sojourn time is computed by the model elements that make up the queuing network. Some components must be computed additionally. To illustrate this, we demonstrate here the sojourn time of a transport unit that travels from station 2 to station 7. For this purpose, we consider the components of the sojourn time and compute their distributions as follows[1]:

> **Waiting time at station** 2: this is the time interval from the arrival of the transport unit at station 2 until the time instant immediately before it is loaded on the vehicle. This is calculated directly by the takted milkrun model.

---

[1] The sojourn time for the material flow from station 4 to station 7 is computed analogously.

Residual loading time at station 2: that is the remaining loading time at the station. For this, we need to compute the number of units to be loaded immediately before the observed transport unit is removed from the station. The loaded quantity at station 2, given by $\overline{X}^2$, is computed by the takted milkrun model element. The distribution, we look for, is the residual of $\overline{X}^2$. Approximating it by the residual lifetime of a renewal process observed immediately after the event occurrence, we obtain the residual distribution.

$$P(R^2 = s) = \frac{1}{E(\overline{X}^2)}(1 - \sum_{v=0}^{s-1} \overline{x}_v^2) \qquad \forall \qquad s = 1, \cdots, K^2 \quad (6.3)$$

where K^2 is the constant limit for the loaded quantity at station 2 (see table A.16). Since the loading time for a transport unit has a unit length, this distribution corresponds also to the distribution of the residual loading time. Note that the residual loading time includes also the loading time of the observed unit.

Driving time to terminal 1: this is an input (see table A.17).

Unloading time at terminal 1: due to unloading time of unit length, this corresponds to the unloaded quantity at terminal 1 and can be computed by the takted milkrun model element.

Loading time at terminal 1: this is an input with a lump sum value of 5 units.

Waiting time at terminal 1: this component is computed by the capacity interval model element.

Driving time to terminal 2: this is also an input (see table A.17).

Loading and unloading at terminal 2: loading and unloading times at terminal 2 are present as inputs. Together, they have a lump-sum value of 10 units.

Waiting time at terminal 2 **for transport to Region** 3: the distribution of this component is computed by the capacitated timeout model.

Driving time to station 6: this component is an input illustrated in table A.17.

Approach	sojourn time st.2 → st.7				
	E	scv	$\sigma_{0.95}$	$\sigma_{0.975}$	$\sigma_{0.99}$
sim	276.880	0.062	395	415	434
analy	272.486	0.063	389	409	430
$\Delta_{rel.}$	0.016	0.012	0.015	0.014	0.009

Approach	sojourn time st.4 → st.7				
	E	scv	$\sigma_{0.95}$	$\sigma_{0.975}$	$\sigma_{0.99}$
sim	293.824	0.058	415	435	457
analy	289.153	0.059	409	430	452
$\Delta_{rel.}$	0.016	0.015	0.014	0.011	0.011

Table 6.1.: Analysis of sojourn times in hub and spoke networks; comparison of analytical results with simulation results

Unloading time at station 6: with equation 6.2, we computed the quantity sorted to region 3. From the transport matrix (see table A.18), we see that 50% of the quantity sorted to region 3 is unloaded at station 6. Similar to equation 6.2, we compute the unloaded quantity at station 6.

Driving time to station 7: this is given in table A.17.

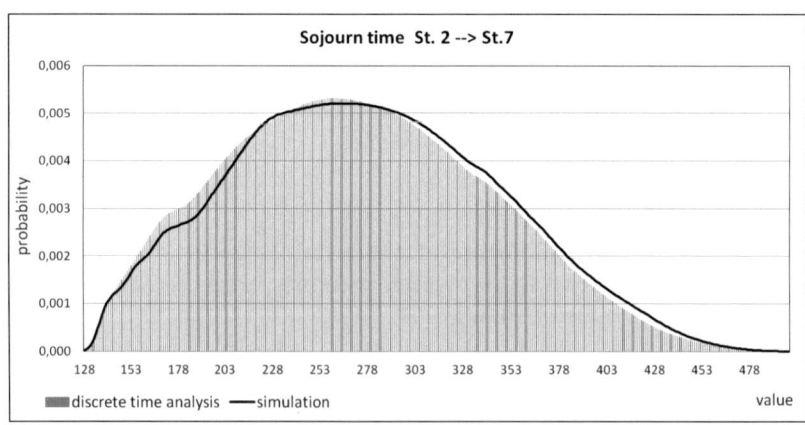

Figure 6.4.: Sojourn time distribution for the flow from station 2 to station 7

Neglecting the dependencies, we convolute these distributions and attain the sojourn time distribution. In table 6.1, we display the mean (E), squared coefficient of variation (scv), 95%, 97.5%, and 99%-quantiles of the sojourn times are listed. Clearly, the relative deviations of the analytical results from simulation results are very low. Furthermore, we compare the results for the sojourn time distribution from station 2 to station 7 with simulation results in figure 6.4 (for the sojourn time from station 4 to station 7, see figure A.6 in section A.4). Although we omit the correlation effects in the analysis, and use some approximate models, the deviations are relatively small. However, it is noteworthy to mention that the results may worsen for larger and complex networks, in which the correlation effect is more pronounced.

7. Conclusion

Consolidated transport processes have been a popular subject-matter of operations research. Numerous optimization problems concerning the route design, scheduling etc. along with their solution methods have been presented in the literature. However, they mostly employ a deterministic approach and neglect stochastic fluctuations in processes. In this work, we were motivated to develop discrete time methods for the stochastic analysis of consolidated transport processes.

The first consolidation strategy, we studied, was inventory consolidation, which is characterized by delaying the transport quantities until a proper shipment size is built. We presented methods for two batch building rules and a batch server queue, which are well-suited to model inventory consolidation strategies.

The first batch building rule, we introduced was capacitated timeout rule. We employed a bivariate discrete Markov chain to derive the joint distribution of the residual time and the remainder. Based on this joint distribution, we were able to derive inter-departure time, departing batch size, and waiting time distributions. The solutions, we presented, are exact, given that inter-arrival time has a discrete distribution. That's why, we investigated the case, in which the inter-arrival time distribution is actually a continuous distribution. We discretized the actual distribution for different values of the unit length (Δt). For this purpose, all the values in the unit length are accumulated and treated as a single discrete value. The results show that the quality of the discrete time approach is determined by the decision on the unit length Δt. For small values of Δt, the results are highly accurate. On the other hand, for high values of Δt, a large interval of values is summarized as one single discrete value. In this way, important information on the characteristics of the inter-arrival time distribution is lost. Hence, the accuracy of the method decreases with increasing Δt values.

The second batch building rule, we modeled, was capacity interval rule. This model is suitable for transport processes, which are initiated as soon as a given minimum utilization for the vehicle is achieved. We assumed that at least one vehicle is available when the minimum utilization is attained. For this rule, we derived also exact solutions for the inter-arrival, departing batch size and the waiting time distributions.

As the batch server, we modeled the $G^X/G^{[L,K]}/1$-queue. In this queue, the transport process can only be initiated, if a specific degree of utilization is reached. Regarding this, it is similar to the capacitated timeout rule. The difference is that the vehicle availability is restricted here. The service time defines the time period, in which a vehicle is not available. For the $G^X/G^{[L,K]}/1$-queue, methods for the inter-departure time, idle time, departing batch, and the waiting time were presented. The methods introduced are exact within an ϵ- neighborhood.

Besides, we demonstrated the application of the capacity interval rule and the $G^X/G^{[L,K]}/1$-queue in a case study. The objective was to evaluate different vehicle dispatching strategies. The strategies, we presented, require a minimum utilization of the employed vehicle to initiate a transport process. However, they differ in the vehicle availability. Firstly, we studied the relation between the mean sojourn time and the vehicle availability offered by different strategies. The results show that the sojourn time is pretty stable for a large range of availability values and then decreases drastically for very low values. Finally, we conducted a cost analysis of the tested strategies, where we differentiated between fixed and variable components of the inventory and the transport costs. For instance, we computed the fixed inventory costs assuming that the storage depot is sized with regard to a given safety level. This would not be possible if we would have employed an analysis based on 2-parameter approximations. To sum up, we gained the following insights throughout our analysis:

- Discrete time batch building processes and queues are well-suited for the analysis of inventory consolidation strategies,

- Decision on Δt should be made wisely, as the results deteriorate for high values of Δt. That's why it is important to choose a Δt-value, which provides a good compromise between accuracy and computation time,

- Under vehicle dispatching strategies, which require a minimum utilization level for the vehicles, it is possible to attain a proper sojourn time with a modest vehicle availability.

Following inventory consolidation strategies, we studied vehicle consolidation. Specifically, we investigated takted and shuttle milkrun systems under general assumptions. Each model is applicable to milkrun systems with arbitrary transport relations. In other words, a station in the system may deliver goods to arbitrary stations in arbitrary proportions and may receive goods from a number of arbitrary stations. The transport relations are summarized in a transport matrix which summarizes the probabilities $p^{i,j}$, i.e. the probability that an arbitrary unit loaded at station i is delivered to station j. We assumed in our analysis that the shipment size at each station is limited. Furthermore, the dependency of the loading (unloading) time required on the quantity loaded (unloaded) is taken into consideration. Employing the polling theory, we computed the queue states, cycle times, tour time, and the waiting time distributions for each milkrun system approximately.

In milkrun systems, many system variables are correlated with each other. First of all, the queue state at a specific queue is correlated with the cycle time, which corresponds to the time interval between two loading processes at the station. When the cycle time is long, the queue length increases. Thus, shipment size gets larger. On the other hand, as the shipment sizes increase, the cycle times get larger. Additionally, the quantities loaded at an arbitrary station and delivered to other stations in the same tour are correlated. For instance, if five transport units are picked at an arbitrary station, the quantity delivered from this station to the successive stations cannot be more than five in this tour. These dependencies do not allow the derivation of the output distributions in closed forms. Therefore, we developed iterative algorithms for both milkrun systems.

In takted milkrun systems, a new tour starts in fixed intervals. In order to cope with the takt time, there may be more than one vehicle in the system. As a simplifying assumption, we have excluded the possibility of an overtake between the vehicles in our analysis. In many practical cases, this assumption is justified, as the chosen takt time is usually long enough to make the occurrence of overtakes very unlikely. For takted milkrun systems, we firstly presented an iterative algorithm, which omits the dependency between the picked and the delivered quantities. The numerical results show that the accuracy of our approach diminishes for the tour and the cycle times when the correlation between the picked and delivered quantities increases. This is the case, when shipments at arbitrary stations have deterministic destinations, thus, $p^{i,j} = \{0,1\}$ for all transport relations. For these cases, we developed an improvement algorithm, which considers also this kind of dependency. This algorithm may be employed to improve the results of the basic algorithm when a highly accurate analysis is sought. Nevertheless, the accuracy of the basic algorithm is sufficient for the early planning phase. Furthermore, we studied the influence of the transport matrix on the performance measures. We analyzed the examples, which differ in their transport matrices. The results show that the influence of the transport matrix on the mean values of the performance measures is minimal. But the distributions, thus, scv, of the performance measure are affected by the transport matrix, especially the tour time and the cycle times. The waiting time seems to show the least sensitivity to the transport matrix.

As the next, we investigated shuttle milkrun systems, in which a vehicle shuttles between the stations. We started the discussion with the question, if cycle times at different stations have the same distributions. Considering the symmetry of the system, it was interesting to find out that the right answer was a "no". This is due to the fact that some delivered quantities originate from the quantity loaded in the actual cycle and some others from the previous cycle. The proportion of the delivered quantity that originates from the actual tour is dependent on the loaded quantity in the actual tour. Depending on the station for which the cycle time is observed, the dependent proportion of the delivered quantities changes. As cycle time is correlated with the sum of all picked and delivered quantities, cycle time distributions are slightly different at different stations. The same conclusion applies for the tour time. Depending on the reference station, where the tour starts and ends, the tour time distributions may look slightly different. For the analysis of the system, we compute the cycle time at station 0 and use it to approximate the cycle times at other stations. We assume in our analysis that a tour starts and ends at station 0. Thus, the tour time corresponds to the cycle time at station 0.

In shuttle milkrun systems, cycle time determines, on one hand, queue states. On the other hand, queue states determine cycle time. That's why, the iterative algorithm is characterized by the computation of queue states based on cycle time distribution from the previous iteration step and the derivation of the cycle time distribution from the computed queue states. As the case with takted milkrun systems, the transport relationships or rather the transport matrix is the most important factor that affects the quality of the results. Although the accuracy may diminish for systems with deterministic transport relations, the results are still very accurate.

Eventually, we investigated terminal consolidation and showed how the models developed in this work can be employed to model hub-and-spoke networks. In this analysis, we omitted the correlation effects. The results are very promising for small networks like the one, we studied. However, for larger networks correlation effects can be accumulated yielding larger deviations from simulation results.

In summary, we made following conclusions from our analyses:

- Due to the correlations between the system variables, iterative algorithms are well-suited to model milkrun systems.

- The most important factor that affects the quality of the iterative algorithms for the takted and shuttle milkrun systems is the transport matrix. For stochastic transport relations, the results are very accurate. For deterministic transport relations, the accuracy of the results is sufficient for the early planning phase.

- Transport matrix has an insignificant effect on the mean values of the performance measures. However, variabilities of the performance measures increase under stochastic transport relations. The increase is significant for the cycle times and the tour time.

- For very long takt times, quantiles of waiting times and queue states increase drastically. On the other hand, mean values show a more linear trend.

- Cycle time distributions at arbitrary stations does not have to be exactly same in the shuttle milkrun system. Similarly, tour time distributions may be different depending on the assumption, at which station a tour starts and ends.

- Introduced models can be used to model hub-and-spoke networks. However, for very large networks, the correlation effects may become more significant.

This work contributed analytical approaches for the analysis of consolidated transport systems. However, there are still some missing models. For instance, we assumed in our models, that transport units or shipments are handled according to the FIFO-Basis. However, this is usually not the case in transport systems, where express deliveries are given the priority. The models, we introduced, can be extended to the assumption of different priority classes. The challenge there is to find the proportion of different classes in the remaining transport units, which were left over in the queue after the start of a transport process (due to capacity restriction). Obviously, lower priority transport units are more likely to be left over in the queue.

We assumed in the analysis of shuttle milkrun systems, that there is only one vehicle that circulates in the route. Nevertheless, there may be multiple vehicles that traverse a closed loop. Moreover, the routes can be so narrow that the vehicles block each other. Hence, discrete time models for closed networks with blocking are needed.

Another problem area is the network analysis. Here, we ignore the correlation effects and use the description of the departure process at a system for the description of the arrival process at the next system. However, the departure process is usually a correlated process. Unfortunately, the majority of the existing models in the literature assumes iid arrival process. Consequently, models are missing that relax the assumption of iid arrival process.

8. Glossary

σ_u	$u\%$ quantile of RV X
$\otimes$	convolution operator
$\Pi_m[x]$	operator which lower bounds distribution x to value m
$\Pi^M[x]$	operator which upper bounds distribution x to value M
$\Delta_m[x]$	operator which shifts distribution x by m units
λ	arrival rate
μ	service rate
ρ	utilization
$\Delta_{rel.}$	relative deviation
A	inter-arrival time
A^i	inter-arrival time at station i
AGV	automated guided vehicle
B	service time
c_X^2	squared coefficient of variation of RV X
C	cycle time
C^i	cycle time at station i, time interval between two successive loading processes at station i
CB	total loading and unloading time in a tour
C_{capa}	fixed costs of a storage place allocated to observation period
C_{FInv}	fixed inventory costs incurred in observation period
C_{FTrans}	fixed transport costs incurred in observation period
CP T	clocked provision strategy with a takt time T
CS^i	cycle segment at station i, total time from the beginning of a tour until the beginning of the loading process at station i
C_{inv}	variable inventory costs per transport unit in observation period
C_{trans}	costs per transport process
$C_{vehicle}$	fixed costs per vehicle assigned to the observation period
C_{VInv}	variable inventory costs incurred in observation period
C_{VTrans}	variable transport costs incurred in observation period
CWL^i	workload contributed by station i
$CWL^{i,I}$	sum of loaded quantity at station i and dependent proportion of quantity delivered from station i in a tour

$CWL^{i,II}$	independent proportion of quantity delivered from station i in a tour
D_{out}	inter-departure time
$E(X)$	expected value (mean) of RV X
G^i	number of transport units arrived within a cycle at station i
iid	independent and identically distributed
h_{max}	maximum idle time
IMT	immediate transport strategy
IT	idle time
K	server capacity
K^i	constant limit for the quantity picked up at station i in a tour
L	minimum batch size
l_{max}	maximum number of arrivals
$l_{m,max}$	maximum number of arrivals in m time units
N	number of arrivals within a collecting process
n_{capa}	number of storage places needed to dimension a depot with a given safety level
n_{trans}	number of transport processes in the observation period
n_{inv}	mean number of transport units in system
$n_{vehicle}$	number of necessary transport vehicles needed for a dispatching strategy
OT	one-way trip time
$p^{j,i}$	probability, that an arbitrary transport unit loaded at station j is delivered to station i
$p(i,m)$	transition probability from an initial remainder of i to m
$p(u,i)(s,j)$	transition probability from residual state (u,i) to (s,j)
$p^1(u,i)(s,j)$	transition probability from residual state (u,i) to (s,j) under case 1 (see section 4.4.1)
$p^2(u,i)(s,j)$	transition probability from residual state (u,i) to (s,j) under case 2 (see section 4.4.1)
$P(C1)$	probability for case 1 (see section 4.4.1)
$P(C2)$	probability for case 2 (see section 4.4.1)
R	residual state, two dimensional variable defining the states of residual inter-arrival time and remainder
R^a	residual inter-arrival time, time interval between the start of a process and the first arrival
R^i	residual inter-arrival time at station i
R^y	remainder, number of remaining transport units due to the limited capacity
r^y_{max}	maximum value of remainder
RT	round-trip time
RV	random variable
S^*	total driving time in a tour

S^i	driving time between station i and the next station
scv	squared coefficient of variation
ST	shuttle transport strategy
T	takt time
T_{obs}	observation time period
t_{out}	maximum collecting time
TT	tour time, duration of a tour
TWL	total workload in a tour, sum of picked and delivered quantities in a tour
$U^{i,j}$	quantity picked up at station i and delivered already to its destination before reaching station j in a tour
$VAR(X)$	variance of RV X
W	waiting time of an arbitrary transport unit
W^i	waiting time of an arbitrary transport unit at station i
W^k	waiting time of an arbitrary unit from the kth batch
$W^{k,l}$	waiting time of an arbitrary unit from the kth batch, given that l arrivals occur in a collecting process
WI^i	sum of independent quantities delivered to station i in a tour
WL^i	workload till station i, sum of picked and delivered quantities until the beginning of loading process at station i
X^i	queue state at station i
$\overline{X}^i$	quantity picked up at station i
X^{+i}	quantity beyond the limit at the start of the loading process at station i
$\widehat{X}^i$	quantity unloaded at station i
$\widehat{X}^{j,i}$	quantity delivered from station j to station i
Y	batch size of an incoming batch
Y_{arr}	number of transport units arrived within a collecting process
Y_{col}	number of transport units collected within a collecting process
Y^i	incoming batch size at station i
Y_{out}	departing batch size
Y_{stor}	number of transport units in storage immediately before a service process

References

Abolnikov, L., J. Dshalalow and A. Dukhovny (1994). First passage processes in Queuing system $M^X/G^r/1$ with service delay discipline. *International Journal of Mathematics and Mathematical Sciences 17*(3), pp. 571–586.

Adan, I., O. Boxma and J. Resing (2001). Queueing models with multiple waiting lines. *Queueing Systems 37*(1), pp. 65–98.

Alfa, A. and Q. He (2008). Algorithmic analysis of the discrete time $GI^X/G^Y/1$ queueing system. *Performance Evaluation 65*(9), pp. 623–640.

Altman, E. and U. Yechiali (1994). Polling in a closed network. *Probability in the Engineering and Informational Sciences 8*(03), pp. 327–343.

Arnold, D. and K. Furmans (2009). *Materialfluss in Logistiksystemen.* Springer.

Arumuganathan, R. and K. Ramaswami (2005). Analysis of a bulk queue with fast and slow service rates and multiple vacations. *Asia-Pacific Journal of Operational Research (APJOR) 22*(2), pp. 239–260.

Baba, Y. (1996). A bulk service $GI/M/1$ queue with service rates depending on service batch size. *Journal of the Operations Research Society of Japan-Keiei Kagaku 39*(1), pp. 25–35.

Bagchi, T. and J. Templeton (1973). Finite waiting space bulk queueing systems. *Journal of Engineering Mathematics 7*(4), pp. 313–317.

Bailey, N. (1954). On queueing processes with bulk service. *Journal of the Royal Statistical Society. Series B (Methodological) 16*(1), pp. 80–87.

Baker, J. and I. Rubin (1987). Polling with a general-service order table. *Communications, IEEE Transactions on 35*(3), pp. 283–288.

Bartholdi, J. and L. Platzman (1989). Decentralized control of automated guided vehicles on a simple loop. *IIE transactions 21*(1), pp. 76–81.

Beekhuizen, P. and J. Resing (2009). Approximation of discrete-time polling systems via structured Markov chains. In: *Proceedings of the Fourth International ICST Conference on Performance Evaluation Methodologies and Tools*, pp. 1–10. ICST (Institute for Computer Sciences, Social-Informatics and Telecommunications Engineering).

Bhat, U. (1964). Imbedded Markov chain analysis of single server bulk queues. *Journal of the Australian Mathematical Society 4*(02), pp. 244–263.

Bitran, G. and D. Tirupati (1989). Approximations for product departures from a single-server station with batch processing in multiproduct queues. *Management science 35*(7), pp. 851–878.

Bolch, G., S. Greiner, H. de Meer and K. Trivedi (1998). *Queueing networks and Markov chains. Modeling and performance evaluation with computer science applications*. Hoboken, NJ: John Wiley & Sons.

Boon, M. and I. Adan (2009). Mixed gated/exhaustive service in a polling model with priorities. *Queueing Systems 63*(1), pp. 383–399.

Boon, M., I. Adan and O. Boxma (2008). A two-queue polling model with two priority levels in the first queue. In: *Proceedings of the 3rd International Conference on Performance Evaluation Methodologies and Tools*, pp. 1–10. ICST (Institute for Computer Sciences, Social-Informatics and Telecommunications Engineering).

Boxma, O., W. Groenendijk and J. Weststrate (1990). A pseudoconservation law for service systems with a polling table. *Communications, IEEE Transactions on 38*(10), pp. 1865–1870.

Boxma, O., H. Levy and U. Yechiali (1992). Cyclic reservation schemes for efficient operation of multiple-queue single-server systems. *Annals of Operations Research 35*(3), pp. 187–208.

Boxma, O. and J. Weststrate (1989). Waiting Times in Polling Systems with Markovian Server Routing. In: *In: G. Stiege and J. S. Lie (Eds.), Messung, Modellierung und Bewertung von Rechensystemen, 5. GI/ITG-Fachtagung*, pp. 89–104. Springer-Verlag.

Bozer, Y., M. Cho and M. Srinivasan (1994). Expected waiting times in single-device trip-based material handling systems. *European Journal of Operational Research 75*(1), pp. 200–216.

Bozer, Y. and M. Srinivasan (1991). Tandem configurations for automated guided vehicle systems and the analysis of single vehicle loops. *IIE transactions 23*(1), pp. 72–82.

Browne, S. and G. Weiss (1992). Dynamic priority rules when polling with multiple parallel servers. *Operations research letters 12*(3), pp. 129–137.

Bruneel, H. and B. Kim (1992). *Discrete-time models for communication systems including ATM*. Kluwer Academic Publishers Norwell, MA, USA.

Burke, P. (1975). Delays in single-server queues with batch input. *Operations Research*, pp. 830–833.

Chakravarthy, S. (1998). Analysis of a priority polling system with group services. *Stochastic Models 14*(1), pp. 25–49.

Chaudhry, M. and U. Gupta (1997). Queue-length and waiting-time distributions of discrete-time $G^x/Geom/1$ queueing systems with early and late arrivals. *Queueing Systems 25*(1), pp. 307–324.

Chaudhry, M. and N. Kim (2003). A complete and simple solution for a discrete-time multi-server queue with bulk arrivals and deterministic service times. *Operations Research Letters 31*(2), pp. 101–107.

Chaudhry, M. and J. Templeton (1983). *A First Course in Bulk Queues*. John Wiley, New York.

Chaudhry, M., J. Templeton and J. Medhi (1992). Computational Results of Multiserver Bulk-Arrival Queues with Constant Service Time $M^x/D/c$. *Operations Research 40*, pp. 229–238.

Choi, B. and K. Park (1992). The M k/m/8 Queue with Heterogeneous Customers in a Batch. *Journal of Applied Probability 29*(2), pp. 477–481.

Cong, T. (1994). On the M X/G/8 Queue with Heterogeneous Customers in a Batch. *Journal of Applied Probability 31*(1), pp. 280–286.

Cooper, R. and G. Murray (1969). Queues served in cyclic order. *Bell Syst. Tech. J 48*(3), pp. 675–689.

Dittmann, R. and F. Hübner (1993). Discrete-Time Analysis of a

Cyclic Service System with Gated Limited Service. Technical report, Lehrstuhl für Informatik, Universität Würzburg.

Dukhovnyy, I. (1979). An approximate model of motion of urban passenger transportation over annular routes. *Engineering Cybernetics 17*(1), pp. 161–162.

Dümmler, M. (1998). Analysis of the departure process of a batch service queueing system. Technical report, Lehrstuhl für Informatik, Universität Würzburg.

Eisenberg, M. (1972). Queues with periodic service and changeover time. *Operations Research 20*(2), pp. 440–451.

Federgruen, A. and Z. Katalan (1996a). Customer waiting-time distributions under base-stock policies in single-facility multi-item production systems. *Naval Research Logistics 43*(4), pp. 533–548.

Federgruen, A. and Z. Katalan (1996b). The stochastic economic lot scheduling problem: cyclical base-stock policies with idle times. *Management Science 42*(6), pp. 783–796.

Federgruen, A. and Z. Katalan (1998). Determining production schedules under base-stock policies in single facility multi-item production systems. *Operations Research 46*(6), pp. 883–898.

Ferguson, M. and Y. Aminetzah (1985). Exact results for nonsymmetric token ring systems. *IEEE transactions on communications 33*(3), pp. 223–231.

Fiems, D., S. De Vuyst and H. Bruneel (2002). The combined gated-exhaustive vacation system in discrete time. *Performance Evaluation 49*(1-4), pp. 227–239.

Fowler, J., N. Phojanamongkolkij, J. Cochran and D. Montgomery (2002). Optimal batching in a wafer fabrication facility using a multiproduct G/G/c model with batch processing. *International Journal of Production Research 40*(2), pp. 275–292.

Fuhrmann, S. (1985). Symmetric queues served in cyclic order. *Operations Research Letters 4*(3), pp. 139–144.

Furmans, K. (2004a). A framework of stochastic finite elements for models of material handling systems. *Progress in material handling research: 8. International Material Handling Research Colloquium, Graz, 2004.*.

Furmans, K. (2004b). Zeitdiskrete Modellierung von Logistikprozessen. Technical report, Institut für Fördertechnik und Logistiksysteme, Universität Karlsruhe.

Gaver, D. (1959). Imbedded Markov chain analysis of a waiting-line process in continuous time. *The Annals of Mathematical Statistics 30*(3), pp. 698–720.

Gnedenko, B. and D. König (1984). *Handbuch der Bedienungstheorie II*, Volume 2. Akademie-Verlag.

Gold, H. and P. Tran-Gia (1993). Performance analysis of a batch service queue arising out of manufacturing system modelling. *Queueing Systems 14*(3), pp. 413–426.

Grasman, S., T. Olsen and J. Birge (2008). Setting basestock levels in multi-product systems with setups and random yield. *IIE Transactions 40*(12), pp. 1158–1170.

Grassmann, W. and J. Jain (1989). Numerical solutions of the waiting time distribution and idle time distribution of the arithmetic GI/G/1 queue. *Operations Research 37*(1), pp. 141–150.

Gupta, S. and J. Goyal (1966). Queues with batch poisson arrivals and hyper-exponential service time distribution. *Metrika 10*(1), pp. 171–178.

Gupta, U. and V. Goswami (2002). Performance analysis of finite buffer discrete-time queue with bulk service. *Computers & Operations Research 29*(10), pp. 1331–1341.

Hashida, O. (1972). Analysis of multiqueue. *Rev. Electrical Communication Laboratories*, pp. 189–199.

Kleinrock, L. and H. Levy (1988). The analysis of random polling systems. *Operations Research 36*(5), pp. 716–732.

Konheim, A. and B. Meister (1974). Waiting lines and times in a system with polling. *Journal of the ACM (JACM) 21*(3), pp. 470–490.

Krieg, G. and H. Kuhn (2002). A decomposition method for multi-product kanban systems with setup times and lost sales. *IIE Transactions 34*(7), pp. 613–625.

Krishna, R., R. Nadarajan and R. Arumuganathan (1998). Analysis of a bulk queue with N-policy multiple vacations and setup times. *Computers & Operations Research 25*(11), pp. 957–967.

Kuehn, P. (1979). Multiqueue systems with nonexhaustive cyclic service. *Bell Syst. Tech. J 58*(3), pp. 671–698.

Lee, D. and B. Sengupta (1992). An approximate analysis of a cyclic server queue with limited service and reservations. *Queueing Systems 11*(1), pp. 153–178.

Lee, H. and M. Srinivasan (1990). The shuttle dispatch problem with compound Poisson arrivals: controls at two terminals. *Queueing Systems 6*(1), pp. 207–221.

Lee, Y. (2001). Discrete-time $Geo^x/G/1$ queue with preemptive resume priority. *Mathematical and Computer Modelling 34*(3-4), pp. 243–250.

Lehoczky, J. (1972). Traffic intersection control and zero-switch queues under conditions of Markov Chain dependence input. *Journal of Applied Probability 9*(2), pp. 382–395.

Leung, K. (1991). Cyclic-service systems with probabilistically-limited service. *IEEE Journal on Selected Areas in Communications 9*(2), pp. 185–193.

Leung, K. (1994). Cyclic-service systems with nonpreemptive, time-limited service. *IEEE Transactions on Communications 42*(8), pp. 2521–2524.

Levy, H. and L. Kleinrock (1991). Polling systems with zero switch-over periods: a general method for analyzing the expected delay. *Performance Evaluation 13*(2), pp. 97–107.

Levy, H. and M. Sidi (1990). Polling systems: Applications, modeling, and optimization. *Communications, IEEE Transactions on 38*(10), pp. 1750–1760.

Levy, H. and M. Sidi (1991). Polling systems with simultaneous arrivals. *Communications, IEEE Transactions on 39*(6), pp. 823–827.

Lye, K. and K. Seah (1992). Random polling scheme with priority. *Electronics Letters 28*(14), pp. 1290–1291.

Mack, C. (1957). The efficiency of N machines uni-directionally patrolled by one operative when walking time is constant and repair times are variable. *Journal of the Royal Statistical Society. Series B (Methodological)*, pp. 173–178.

Mack, C., T. Murphy and N. Webb (1957). The efficiency of N machines uni-directionally patrolled by one operative when walking time and repair times are constants. *Journal of the Royal Statistical Society. Series B (Methodological) 19*(1), pp. 166–172.

Madan, K., A. Al-Nasser and A. Al-Masri (2004). On queue with optional re-service. *Applied Mathematics and Computation 152*(1), pp. 71–88.

Meng, G. and S. Heragu (2004). Batch size modeling in a multi-item, discrete manufacturing system via an open queuing network. *IIE transactions 36*(8), pp. 743–753.

Neuts, M. (1965). The busy period of a queue with batch service. *Operations Research 13*(5), pp. 815–819.

Neuts, M. (1967). A general class of bulk queues with Poisson input. *The Annals of Mathematical Statistics 38*(3), pp. 759–770.

Newell, G. (1998). The rolling horizon scheme of traffic signal control. *Transportation Research Part A: Policy and Practice 32*(1), pp. 39–44.

Olsen, T. and R. Van der Mei (2003). Polling systems with periodic server routing in heavy traffic: distribution of the delay. *Journal of Applied Probability 40*(2), pp. 305–326.

Powell, W. and P. Humblet (1986). The bulk service queue with a general control strategy: Theoretical analysis and a new computational procedure. *Operations Research 34*(2), pp. 267–275.

Prabhu, N. (1987). Stochastic comparisons for bulk queues. *Queueing Systems 1*(3), pp. 265–277.

Schleyer, M. (2007). *Discrete time analysis of batch processes in material flow systems*. Dissertation, Institut für Fördertechnik und Logistiksysteme (IFL), Universität Karlsruhe.

Schleyer, M. and K. Furmans (2007). An analytical method for the calculation of the waiting time distribution of a discrete time $G/G/1$-queueing system with batch arrivals. *OR Spectrum 29*(4), pp. 745–763.

Sim, S. and J. Templeton (1983). Computational procedures for steady-state characteristics of unscheduled multi-carrier shuttle systems. *European Journal of Operational Research 12*(2), pp. 190–202.

Simão, H. and W. Powell (1988). Waiting time distributions for transient bulk queues with general vehicle dispatching strategies. *Naval Research Logistics 35*(2), pp. 285–306.

Srinivasan, M., H. Levy and A. Konheim (1996). The individual station technique for the analysis of cyclic polling systems. *Naval Research Logistics 43*(1), pp. 79–101.

Takagi, H. (1986). *Analysis of polling systems*. Cambridge, MA: M.I.T. Press.

Takagi, H. (1988). Queuing analysis of polling models. *ACM Computing Surveys (CSUR) 20*(1), pp. 5–28.

Takagi, H. (1991). Analysis of finite-capacity polling systems. *Advances in Applied Probability 23*(2), pp. 373–387.

Takagi, H. (2000). Analysis and application of polling models. *Performance Evaluation: Origins and Directions*, pp. 423–442.

Takagi, H. and K. Leung (1994). Analysis of a discrete-time queueing system with time-limited service. *Queueing Systems 18*(1), pp. 183–197.

Takahashi, Y. and B. Kumar (1995). Pseudo-conservation law for a priority polling system with mixed service strategies. *Performance Evaluation 23*(2), pp. 107–120.

Tran-Gia, P. (1993). Discrete-time analysis technique and application to usage parameter control modelling in ATM systems. In: *8th Australian Teletraffic Research Seminar*.

Tran-Gia, P. (1996). *Analytische Leistungsbewertung verteilter Systeme*. Springer.

Tran-Gia, P. (2002). Analysis of polling systems with general input process and finite capacity. *IEEE Transactions on Communications 40*(2), pp. 337–344.

Tran-Gia, P. and H. Ahmadi (1988). Analysis of a discrete-time $G^x/D/1$-S queueing system with applications in packet-switching systems. In: *IEEE INFOCOM*, Volume 88, pp. 0861–0870.

Tran-Gia, P. and T. Raith (1988). Performance analysis of finite capacity polling systems with nonexhaustive service. *Performance Evaluation 9*(1), pp. 1–16.

Tran-Gia, P. and A. Schömig (1996). Discrete-time analysis of batch servers with bounded idle time. In: *In Modelling and Simulation 1996.(Proceedings of the European Simulation Multiconference)*. Citeseer.

Van der Mei, R. and S. Borst (1997). Analysis of multiple-server polling systems by means of the power-series algorithm. *Stochastic Models 13*(2), pp. 339–369.

Van Ommeren, J. (1990). Simple Approximations for the Batch-Arrival $M^x/G/1$ Queue. *Operations Research 38*(4), pp. 678–685.

Vishnevskii, V. and O. Semenova (2006). Mathematical methods to study the polling systems. *Automation and Remote Control 67*(2), pp. 173–220.

Whitt, W. (1993). Approximations for the GI/G/m queue. *Production and Operations Management 2*(2), pp. 114–161.

Winands, E., I. Adan and G. Van Houtum (2006). Mean value analysis for polling systems. *Queueing Systems 54*(1), pp. 35–44.

Yao, D., M. Chaudhry and J. Templeton (1984). Analyzing the Steady-State Queue $GI^x/G/1$. *Journal of the Operational Research Society 35*(11), pp. 1027–1030.

Zhao, Y. and L. Campbell (1996). Equilibrium probability calculations for a discrete-time bulk queue model. *Queueing Systems 22*(1), pp. 189–198.

A. Appendix

A.1. Analysis of the $G^X/G^{[L,K]}/1$-queue

Inter-arrival Time		Batch Size		Service Time	
i	a_i	i	y_i	i	b_i
9	0.25	6	0.1	20	0.23
10	0.25	7	0.2	22	0.66
11	0.32	8	0.1	24	0.11
14	0.18	9	0.1	Parameter	
		10	0.3	L	15
		11	0.1	K	25
		12	0.1		

Table A.1.: System under consideration for the analysis of the $G^X/G^{[L,K]}/1$-queue in section 4.4.5

System figures	
$C_{vehicle}$	60000
C_{capa}	5000
C_{inv}	200
C_{trans}	400
T_{obs}	38400

Table A.2.: Figures for the cost analysis in section 4.4.5

Inter-arrival Time		Batch Size		One-way Trip		Round Trip	
i	a_i	i	y_i	i	ot_i	i	rt_i
4	0.27	3	0.1	6	0.4	13	0.1
5	0.13	4	0.4	9	0.5	18	0.6
7	0.33	5	0.4	12	0.1	23	0.3
8	0.17	7	0.1				
9	0.1						

Table A.3.: System under consideration in section 4.4.5

i	Simulation		Analytical Approach			
	lower bound	upper bound	$r^y_{max}:5$	$r^y_{max}:$ 11,22,30	$r^y_{max}:5$	$r^y_{max}:$ 11,22,30
15	0.05688	0.05787	0.05878	0.05747	nok	ok
16	0.10371	0.10473	0.10445	0.10418	ok	ok
17	0.15050	0.15164	0.15181	0.15141	nok	ok
18	0.12478	0.12655	0.12526	0.12525	ok	ok
19	0.12067	0.12190	0.12108	0.12099	ok	ok
20	0.13853	0.14036	0.13934	0.13928	ok	ok
21	0.09423	0.09542	0.09518	0.09509	ok	ok
22	0.08126	0.08210	0.08206	0.08198	ok	ok
23	0.04123	0.04243	0.04147	0.04168	ok	ok
24	0.03585	0.03632	0.03593	0.03620	ok	ok
25	0.04628	0.04675	0.04464	0.04649	nok	ok

Table A.4.: Validation of the departing batch size distribution for the $G^X/G^{[L,K]}/1$-queue in section 4.4.5

Note that nok: the probability value computed by the analytical approach lies outside the confidence interval, ok: the probability value computed by the analytical approach lies in the confidence interval.

i	Simulation		Analytical Approach			
	lower bound	upper bound	$r^y_{max}:5$	$r^y_{max}:$ 11,22,30	$r^y_{max}:5$	$r^y_{max}:$ 11,22,30
20	0.16565	0.16655	0.16594	0.16653	ok	ok
21	0.01367	0.01428	0.01402	0.01391	ok	ok
22	0.48810	0.49012	0.48678	0.48837	nok	ok
23	0.04916	0.05029	0.05033	0.04985	ok	ok
24	0.11385	0.11461	0.11413	0.11411	ok	ok
25	0.03896	0.03980	0.03998	0.03938	ok	ok
26	0.02084	0.02148	0.02149	0.02125	ok	ok
27	0.01872	0.01993	0.01970	0.01949	ok	ok
28	0.01556	0.01617	0.01620	0.01594	ok	ok
29	0.01128	0.01183	0.01166	0.01164	ok	ok
30	0.01234	0.01262	0.01263	0.01251	ok	ok
31	0.00962	0.00994	0.00981	0.00980	ok	ok
≥32	0.03684	0.03779	0.03733	0.03722	ok	ok

Table A.5.: Validation of the inter-departure time distribution for the $G^X/G^{[L,K]}/1$-queue in section 4.4.5

Note that nok: the probability value computed by the analytical approach lies outside the confidence interval, ok: the probability value computed by the analytical approach lies in the confidence interval.

i	Simulation		Analytical Approach			
	lower bound	upper bound	$r^y_{max}:5$	$r^y_{max}:$ 11,22,30	$r^y_{max}:5$	$r^y_{max}:$ 11,22,30
0	0.17153	0.17279	0.17339	0.17222	ok	ok
1	0.04566	0.04610	0.04577	0.04592	nok	ok
2	0.05288	0.05345	0.05290	0.05316	ok	ok
3	0.04601	0.04647	0.04582	0.04613	nok	ok
4	0.04153	0.04200	0.04156	0.04186	ok	ok
5	0.03241	0.03270	0.03230	0.03251	nok	ok
6	0.02739	0.02776	0.02763	0.02749	ok	ok
7	0.02167	0.02234	0.02194	0.02203	ok	ok
8	0.01693	0.01740	0.01799	0.01718	nok	ok
9	0.03364	0.03393	0.03392	0.03377	ok	ok
10	0.04169	0.04216	0.04232	0.04194	nok	ok
11	0.08446	0.08541	0.08517	0.08500	ok	ok
12	0.04352	0.04406	0.04350	0.04384	nok	ok
13	0.05132	0.05179	0.05104	0.05146	nok	ok
14	0.08934	0.08997	0.08952	0.08972	ok	ok
15	0.03870	0.03900	0.03857	0.03892	nok	ok
16	0.03012	0.03056	0.03010	0.03027	nok	ok
17	0.02541	0.02582	0.02546	0.02556	ok	ok
18	0.02246	0.02274	0.02256	0.02259	ok	ok
19	0.02043	0.02106	0.02090	0.02078	ok	ok
20	0.01959	0.01979	0.01983	0.01979	nok	ok
21	0.01472	0.01494	0.01491	0.01483	ok	ok
≥22	0.02292	0.02345	0.02291	0.02305	nok	ok

Table A.6.: Validation of the waiting time distribution for the $G^X/G^{[L,K]}/1$-queue in section 4.4.5

Note that nok: the probability value computed by the analytical approach lies outside the confidence interval, ok: the probability value computed by the analytical approach lies in the confidence interval.

A.2. Analysis of the takted milkrun systems

Inter-arrival time distributions					Driving time distributions				
Station	0	1	2	3	Station	0	1	2	3
i	a_i	a_i	a_i	a_i	i	s_i	s_i	s_i	s_i
20	0.2	0.5	0.1	0.3	14	0	0	0	0.4
21	0.7	0.5	0	0.4	15	0	0	1	0.6
22	0.1	0	0.1	0.3	16	0	0	0	0
23	0	0	0	0	17	0	0	0	0
24	0	0	0.8	0	18	0	0.5	0	0
25	0	0	0	0	19	0	0	0	0
					20	1	0.5	0	0
Batch size distributions					Constant limits				
Station	0	1	2	3	Station	0	1	2	3
i	y_i	y_i	y_i	y_i	K^i	3	4	4	3
1	0.9	0.6	0.5	1					
2	0.1	0.4	0.5	0					

Table A.7.: Inter-arrival, driving time, batch size distributions, and constant limits for example 1 and example 2 analyzed in section 5.1.9. Note that service time is one ($b_1 = 1$) and the takt time is 42 time units in the given examples.

from\to	St. 0	St. 1	St. 2	St. 3
St. 0	0	1	0	0
St. 1	1	0	0	0
St. 2	1	0	0	0
St. 3	1	0	0	0

Table A.8.: Transport matrix for example 1 in section 5.1.9

from\to	St. 0	St. 1	St. 2	St. 3
St. 0	0	0.9	0.1	0
St. 1	0	0	0.5	0.5
St. 2	0.7	0.3	0	0
St. 3	1	0	0	0

Table A.9.: Transport matrix for example 2 in section 5.1.9

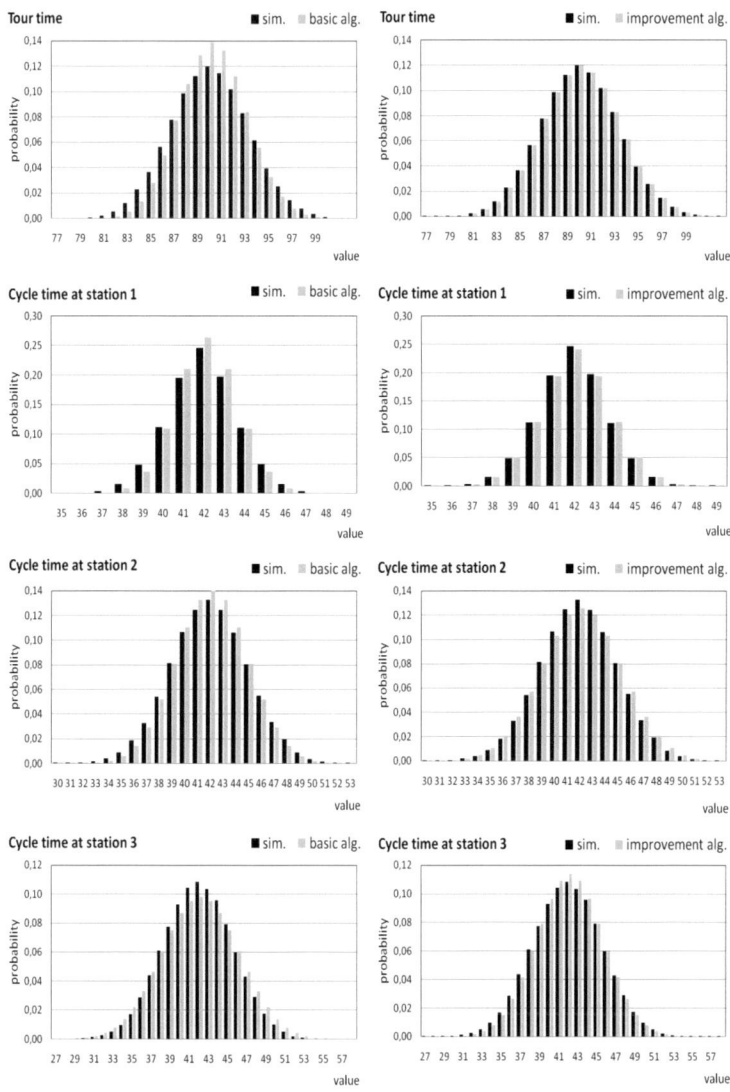

Figure A.1.: Analysis of takted milkrun systems: tour time and cycle time distributions attained by the basic and the improvement algorithms in comparison to simulation results for example 2 in section 5.1.9

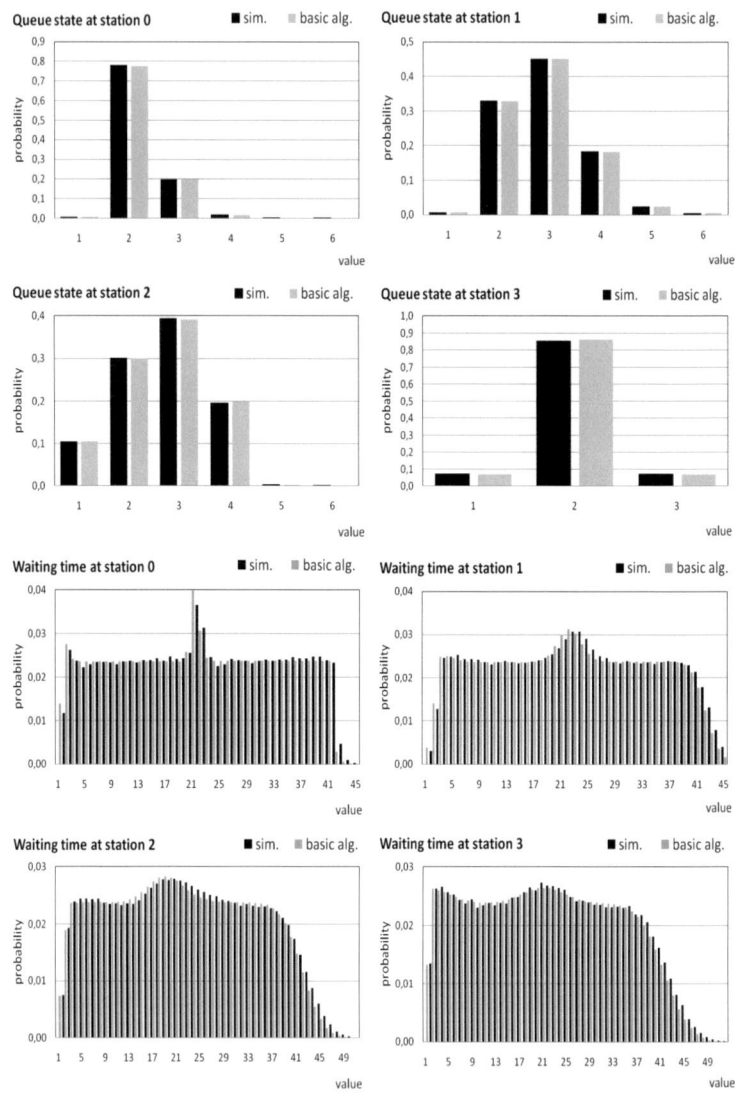

Figure A.2.: Analysis of takted milkrun systems: queue state and waiting time distributions attained by the basic algorithm in comparison to simulation results for example 2 in section 5.1.9

A.3. Analysis of the shuttle milkrun systems

Inter-arrival time distributions				Batch size distributions			
Station	0	1	2	Station	0	1	2
i	a_i	a_i	a_i	i	y_i	y_i	y_i
12	0.5	0.2	0	1	0.9	0.9	0.3
13	0.4	0	0	2	0.05	0.1	0.4
14	0.1	0.7	0	3	0.05	0	0.3
15	0	0.1	0	Driving time distributions			
16	0	0	0	Station	0	1	2
17	0	0	0	i	s_i	s_i	s_i
18	0	0	0	5	0.1	0.4	0.2
19	0	0	0	6	0.8	0.6	0.5
20	0	0	0	7	0.1		0.3
21	0	0	0				
22	0	0	0	Constant limits		Service time dist.	
23	0	0	0	Station	K^i	i	b_i
24	0	0	0	0	4	1	0.9
25	0	0	0.2	1	4	2	0.1
26	0	0	0.6	2	4		
27	0	0	0				
28	0	0	0.2				

Table A.10.: Inter-arrival, driving time, batch size, service time distributions, and constant limits for example 1 analyzed in section 5.2.6.

from \to	St. 0	St. 1	St. 2
St. 0	0	1	0
St. 1	0	0	1
St. 2	0	1	0

from\to	St. 0	St. 1	St. 2	St. 3	St. 4
St. 0	0	0.4	0	0.6	0
St. 1	1	0	0	0	0
St. 2	0.3	0.2	0	0.4	0.1
St. 3	0	0.5	0.4	0	0.1
St. 4	0.8	0	0	0.2	0

Table A.11.: Transport matrix for example 1 in section 5.2.6

Table A.12.: Transport matrix for example 2 in section 5.2.6

Inter-arrival time distributions						Driving time distributions					
Station	0	1	2	3	4	Station	0	1	2	3	4
i	a_i	a_i	a_i	a_i	a_i	i	s_i	s_i	s_i	s_i	s_i
15	0.5	0.8	0	0	0	9	0	0	0	1	0
16	0	0	0	0	0	10	0.5	0	0	0	0
17	0	0	0	0	0	11	0	0	0	0	0
18	0	0	0	0	0	12	0	0.6	0	0	0
19	0	0	0	0	0	13	0	0	0	0	0
20	0.5	0.2	0	0	0	14	0	0.4	0	0	0
21	0	0	0	0	0	15	0.5	0	0.9	0	0
22	0	0	0	0	0	16	0	0	0.1	0	0
23	0	0	0	0	0	17	0	0	0	0	0
24	0	0	0	0	0	18	0	0	0	0	0
25	0	0	0.1	1	0.65	19	0	0	0	0	0
26	0	0	0	0	0	20	0	0	0	0	0
27	0	0	0	0	0	21	0	0	0	0	0
28	0	0	0	0	0	22	0	0	0	0	0
29	0	0	0	0	0	23	0	0	0	0	0.85
30	0	0	0.9	0	0.35	24	0	0	0	0	0.15

Batch size distributions						Constant limits				Service time	
Station	0	1	2	3	4	Station	Station	Station	K^i	i	b_i
i	y_i	y_i	y_i	y_i	y_i	0	15	3	16	1	1
1	0.75	0.95	1		0.5	1	14	4	12		
2	0.25	0.05		1	0.5	2	8				

Table A.13.: Inter-arrival, driving time, batch size, service time distributions, and constant limits for example 2 analyzed in section 5.2.6.

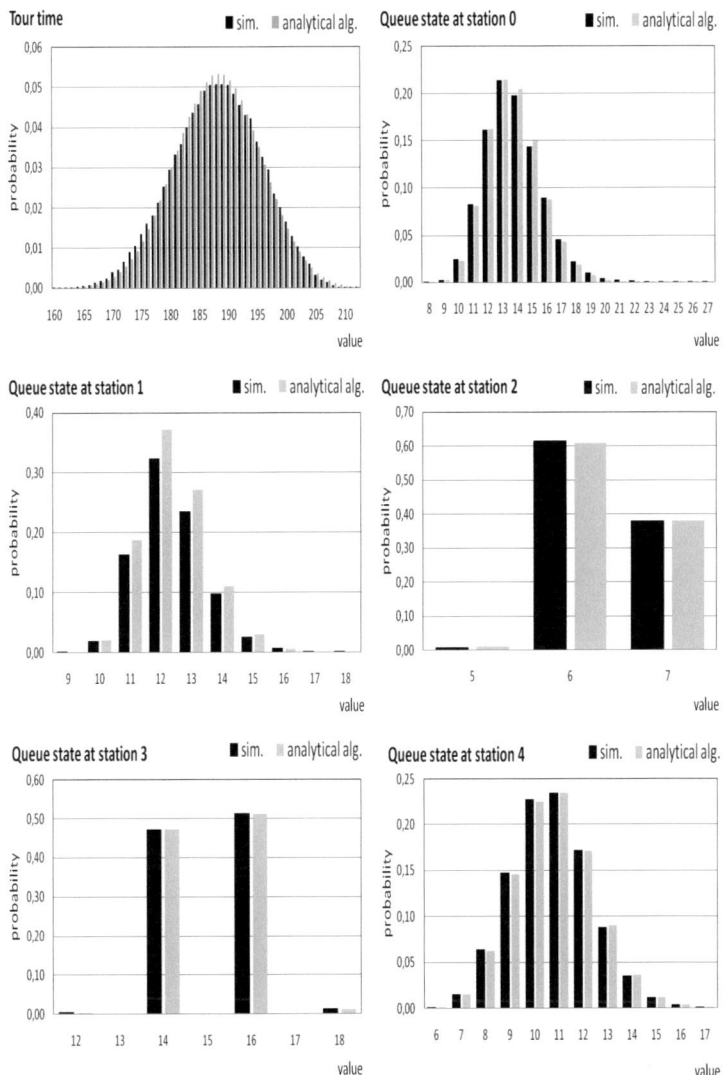

Figure A.3.: Analysis of shuttle milkrun systems: tour time and queue state distributions attained by the analytical algorithm in comparison to simulation results for example 2 in section 5.2.6

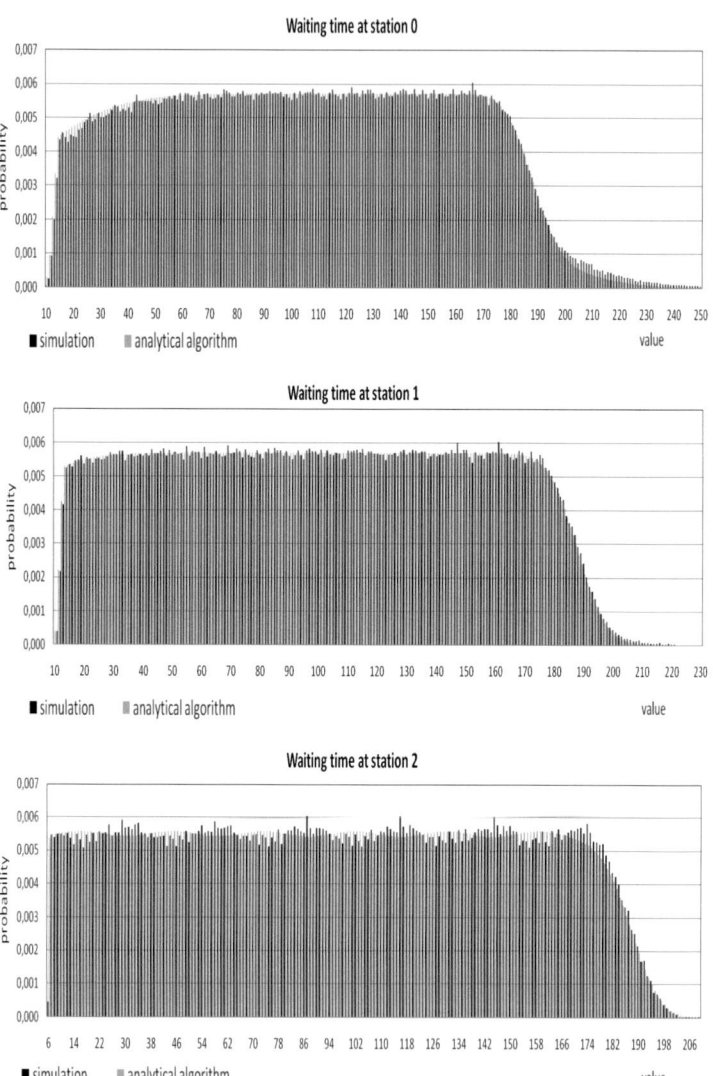

Figure A.4.: Analysis of shuttle milkrun systems: waiting time distributions attained by the analytical algorithm in comparison to simulation results for stations 0, 1, and 2 in example 2 in section 5.2.6

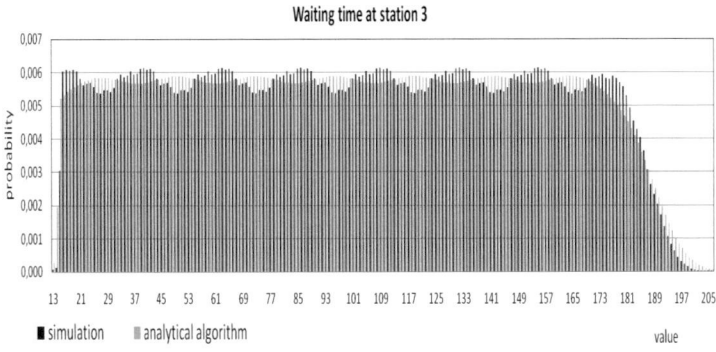

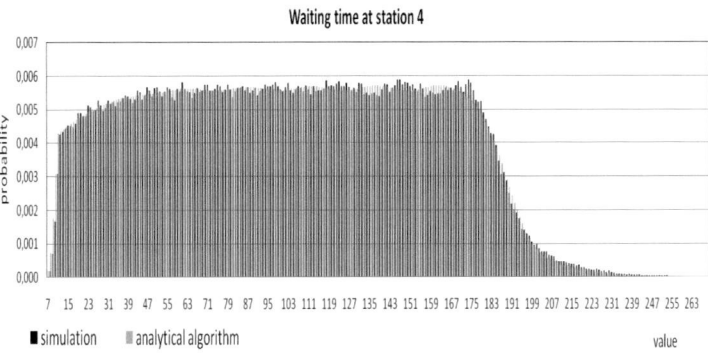

Figure A.5.: Analysis of shuttle milkrun systems: waiting time distributions attained by the analytical algorithm in comparison to simulation results for stations 3 and 4 in example 2 in section 5.2.6

A.4. Numerical case study

	Inter-arrival time distributions							
	Terminal 1 (Reg. 1)	Terminal 1 (Reg 2)	St. 1	St. 2	St. 3	St. 4	St. 5	St. 7
i	a_i	a_i	a_i	a_i	a_i	a_i	a_i	a_i
9	0	0	0.1	0	0	0	0	0
10	0.1	0	0	0	0	0	0	0
11	0.2	0	0.3	0	0	0	0	0
12	0.4	0	0	0.1	0	0	0	0
13	0.2	0	0.6	0.3	0	0	0	0
14	0.1	0	0	0.1	0	0	0	0
15	0	0.5	0	0.5	0	0	0	0
16	0	0	0	0	0	0	0.75	0.2
17	0	0	0	0	0	0	0	0.8
18	0	0	0	0	0.4	0	0.25	0
19	0	0	0	0	0	0	0	0
20	0	0.5	0	0	0.6	0.2	0	0
21	0	0	0	0	0	0.8	0	0

Table A.14.: Inter-arrival time distributions for the numerical case in section 6.2. Note that Terminal 1 (Reg. 1) stands for the inter-arrival of transport units that are transferred from terminal 1 to region 1

	Batch size distributions							
	Terminal 1 (Reg. 1)	Terminal 1 (Reg 2)	St. 1	St. 2	St. 3	St. 4	St. 5	St. 7
i	y_i	y_i	y_i	y_i	y_i	y_i	y_i	y_i
1	0.7	1	1	0.5	0.9	1	0.85	1
2	0.1	0	0	0.4	0.1	0	0.15	0
3	0.1	0	0	0.1	0	0	0	0
4	0.05	0	0	0	0	0	0	0
5	0.05	0	0	0	0	0	0	0

Table A.15.: Batch size distributions for the numerical case in section 6.2. Note that Terminal 1 (Reg. 1) stands for the batch size of transport units that are transferred from terminal 1 to region 1

Node	K^i	Node	K^i
Terminal 1 (Reg. 1)	7	Station 4	5
Terminal 1 (Reg. 2)	5	Station 5	6
Station 1	5	Station 6	0
Station 2	6	Station 7	20
Station 3	5		

Table A.16.: Constant limit for the collection quantities for the numerical case in section 6.2. Note that K^i for Terminal 1 (Reg. 1) stands for the maximum quantity that is loaded at terminal 1 and distributed in region 1

Driving time distributions							
start	Ter. 1	St. 3	St. 4	St. 5	Ter. 2	St. 6	St. 7
end	St. 3	St. 4	St. 5	Ter. 1	St. 6	St. 7	Ter. 2
i	s_i	s_i	s_i	s_i	s_i	s_i	s_i
5	0.7	0	0	0	0	0	0
6	0	0	0	0	0	0	0.3
7	0	0.5	0	0.45	0	0	0
8	0	0	0	0	0	0	0
9	0	0	0	0	0	0	0.4
10	0	0	0.5	0	0.5	0	0
11	0	0	0	0	0	0.9	0
12	0	0.5	0	0.55	0.5	0.1	0
13	0	0	0	0	0	0	0.3
14	0	0	0	0	0	0	0
15	0.3	0	0.5	0	0	0	0
16	0	0	0	0	0	0	0

start	Ter. 1	St. 1	St. 2	Ter. 1
end	St. 1	St. 2	Ter. 1	Ter. 2
i	s_i	s_i	s_i	s_i
20	0.4	0	0.2	0
21	0.3	0	0	0
22	0.3	0	0	0
23	0	0	0.8	0
24	0	0	0	0
25	0	0	0	0
26	0	0	0	0
27	0	0	0	0
28	0	0.9	0	0
29	0	0	0	0
30	0	0.1	0	0
50	0	0	0	1

Table A.17.: Driving time distributions for the numerical case in section 6.2

from \to	Ter. 1	Ter. 2	St.1	St.2	St.3	St.4	St.5	Reg. 3	St. 6	St.7
Ter. 1	0	1	0.5	0.5	0.5	0.4	0.1	0	0	0
Ter. 2	0	0	0	0	0	0	0	0.3	0	0
St.1	0.7	0	0	0.3	0	0	0	0	0	0
St.2	0.8	0	0.2	0	0	0	0	0	0	0
St.3	0.4	0	0	0	0	0.3	0.3	0	0	0
St.4	0.8	0	0	0	0.2	0	0	0	0	0
St.5	0.6	0	0	0	0.3	0.1	0	0	0	0
Reg. 3	0	1	0	0	0	0	0	0	0.5	0.5
St.7	0	1	0	0	0	0	0	0	0	0

Table A.18.: Transport matrix for the numerical case in section 6.2

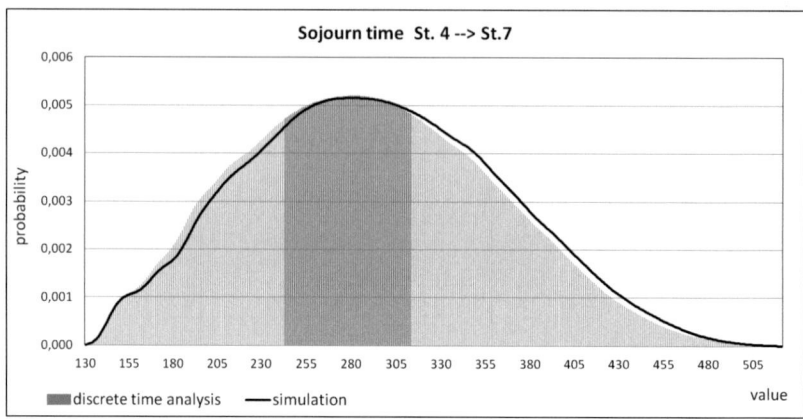

Figure A.6.: Sojourn time distribution for the flow from station 4 to station 7 displayed in section 6.2